Praise for *The Ecological Gardener*

'Engaging and quirky; full of ideas and ins[illegible] garden projects that you'll be itching to try for yourself.'

—**DAVE GOULSON**, author of *The Garden Jungle*

'Finally, a book for UK gardeners who recognise the desperate need to share their gardens with nature. In *The Ecological Gardener,* Matt Rees-Warren explains why every square inch of planet Earth, including our gardens, has ecological significance, and he tells us exactly how to increase that significance in ways that will benefit us all. Excellent, timely, essential!'

—**DOUGLAS W. TALLAMY**, author of *Nature's Best Hope*

'Matt Rees-Warren has distilled his experience, observation and passion for the natural world into an incredibly informative narrative, with practical examples to help us tread lightly on the land. Gardens are for people, but in our endeavour to create beauty, we forget that our space is shared with cohabitants and time. Matt takes us through the garden with an ecological lens – just what we need to view a future that is in harmony with nature itself.'

—**ARIT ANDERSON**, garden designer, chartered member of The Landscape Institute

'Gardening for nature shouldn't be a radical act, but it takes courage to trade power tools and pesticides for hand tools and native plants. *The Ecological Gardener* helps you brave the rewilding of your own patch through practical, inspirational advice for growing lightly on the land. With Matt Rees-Warren and the natural world as your guides, you can welcome all your wild neighbors, from slugs to the newts who eat them, to your life-sustaining space.'

—**NANCY LAWSON**, author of *The Humane Gardener*

‘Gardening fads and styles come and go, from topiaries to green walls to Tropicalisimo. Yet the arc of understanding what a garden is slowly bends to the less controlled and naturally friendly. As our understanding of the connectedness and mostly concealed weave and warp of “life entire” gains steam, so too does our desire to fully embrace the whole. Within this context, Matt Rees-Warren does an admirable job in explaining the objectives while providing the garden-maker a satisfyingly readable map from which to find one’s personal paradise that welcomes the natural world into our own backyard.’

—**DANIEL J. HINKLEY**, author of *Windcliff* and *The Explorer’s Garden*

The Ecological Gardener

HOW TO CREATE BEAUTY AND BIODIVERSITY FROM THE SOIL UP

MATT REES-WARREN

Chelsea Green Publishing
White River Junction, Vermont
London, UK

Project Manager: Natalie T. Jones
Commissioning Editor: Jonathan Rae
Developmental Editor: Michael Metivier
Copy Editor: Lisa Morris
Proofreader: Anne Sheasby
Indexer: Shana Milkie
Designer: Melissa Jacobson

Printed in the United States of America.
First printing March 2021.
10 9 8 7 6 5 4 3 2 1 21 22 23 24 25

Our Commitment to Green Publishing
Chelsea Green sees publishing as a tool for cultural change and ecological stewardship. We strive to align our book manufacturing practices with our editorial mission and to reduce the impact of our business enterprise in the environment. We print our books and catalogues on chlorine-free recycled paper, using vegetable-based inks whenever possible. This book may cost slightly more because it was printed on paper that contains recycled fiber, and we hope you'll agree that it's worth it. *The Ecological Gardener* was printed on paper supplied by Versa that is made of recycled materials and other controlled sources.

Library of Congress Cataloging-in-Publication Data
Names: Rees-Warren, Matt, author.
Title: The ecological gardener : how to create beauty and biodiversity from the soil up / Matt Rees-Warren.
Description: White River Junction, VT : Chelsea Green Publishing, 2021. |
Includes bibliographical references and index.
Identifiers: LCCN 2020055948 (print) | LCCN 2020055949 (ebook) | ISBN 9781645020073 (paperback) | ISBN 9781645020080 (ebook)
Subjects: LCSH: Organic gardening. | Gardening—Environmental aspects.
Classification: LCC SB453.5 .R42 2021 (print) | LCC SB453.5 (ebook) | DDC 635/.0484—dc23
LC record available at https://lccn.loc.gov/2020055948
LC ebook record available at https://lccn.loc.gov/2020055949

Chelsea Green Publishing
85 North Main Street, Suite 120
White River Junction, Vermont USA

Somerset House
London, UK

www.chelseagreen.com

For my wife and daughters

CONTENTS

Introduction

Over the many years I've worked in gardens, I've worn many hats – both literal and figurative! I've broken ground in the coldest depths of a UK winter, and studied the karri tree in the blistering heat of the Western Australian bush. I've led vast redesigning projects in the public realm and spent many hours propagating seedlings in dilapidated greenhouses of quiet country estates. Labels and titles mean little to me, but I've been given many: landscaper, plant buyer, gardener, horticulturist, student of botany, designer, head gardener, nursery manager, veg grower and garden writer. I spend my days in the muck and the flowers, with all the trials and tribulations and joys and satisfaction that go with a weathered life. Gardening is in my bones; not only has it provided for me and my family, but it has also taught me more than I ever thought I'd gain from a simple and honest occupation.

When you work with your hands in the soil and the sun on your back, you take in the natural world at close quarters, in an immersive way. You go where others do not, and see what they do not see. You can glimpse wildlife – which is often illusive – more readily and in more detail, understand soil health with a greater depth of wisdom, and more closely observe the way water moves and collects through the ground. There is an almost overwhelming amount of knowledge to glean from nature. The study of plants alone can occupy many years of a gardener's lifetime.

For me, however, this tactile, tangible closeness with nature has also revealed gardens as bellwethers of a changing landscape. I have seen first-hand the damaging effects non-organic lawn fertilisation can have on the watercourses of a garden, and noticed the lifelessness of a relentlessly cultivated flower. I've witnessed the folly of rigidly aesthetical design unfold and been privy to the exhaustive lengths of control we seek to bring over nature's wild habit, all while the seasons and prevailing climate have changed and become more dynamic and extreme, with climate records tumbling by the year, not by the decade.

As a civilisation we have imprinted on the order of the natural world so dramatically as to have changed its composition beyond what is sustainable, and we now find ourselves at a crossroads where, if we do not choose a new path forward, we face a future of even greater extremes and disharmony. And yet we hold within our grasp the possibility of readdress and regeneration, the chance to reset the clock for ourselves and for many generations to come.

Reimagining how we garden may seem like a small way to help mitigate our ecological crisis, but it's an important one, and it has the ability to make a substantial difference. Gardens are natural spaces after all, the great green lungs of every nation; individual and idiosyncratic but together fundamental to their local and wider ecosystems. How we guide them will be the difference between a future in which they play a key role in restoring nature's health, or one in which we continue our parsimonious approach to nature's needs. While there needs to be a collective changing of the guard, from the community scale to the planetary, small acorns make mighty oaks, and it's also what we do as individual gardeners, today and tomorrow, that matters.

To garden is to foster a bond between ourselves and the natural world.

Indeed, much has changed for the better since I first began gardening. I now design and maintain more wildflower meadows than ever before, and I've found more clients can be encouraged to make their own compost and leave areas of their gardens a little rough and wild. My clients also think twice about easy-fix pesticide remedies, and make tangible efforts to harvest rainwater rather than using mains water. However, the pace of change has been slow, and the uptake of ideas selective and restrained, despite growing collective consciousness about alternatives to conventional gardening. I understand this more than most, as the perils of the garden owner are presented to me on a daily basis. For my clients, I'm a reference through a horticultural world full of mysteries, helping to ease what can be a daunting and confusing journey towards more regenerative practices.

I intend *The Ecological Gardener* to fulfil a similar role for all those who leaf through its pages, whether seasoned and green-fingered or a wide-eyed novice. I hope to guide and inspire in equal measure, working towards a new ethos, a changed method and a fresh approach to the gardening craft. This book places ecological considerations above all else, but without reducing the importance of gardens as platforms for individual expression, aesthetical value and style. Many traditional approaches to gardening need to be reconsidered, and many more new and fresh ideas need to be embraced.

The Ecological Gardener deals with the more aesthetical and ecological aspects of gardens and gardening rather than vegetable production. While growing food and increasing biodiversity are not mutually exclusive, almost all vegetables grown in the UK are not native and this changes the way wildlife and ecosystems exist (or don't exist) in our gardens. To me, this fact marks a dividing line between those gardens whose primary aim is to encourage a wide and diverse ecology of wildlife and those that prioritise production and human sustenance above all. In an ecology-first garden, a blackberry plant covered in aphids is a sign of abundance, not lack of production or yield. However, the ecological gardening mentality does not defend plants at all costs. In fact, it even avoids many organic practices such as biological control and organic sprays. Growing food organically is hugely rewarding and possibly our greatest route to sustainable sustenance, but for balance we also need gardens that promote wildlife and beauty.

Throughout this book these ideas are given room for their promise to be realised, with their methods laid out step-by-step so they can be applied in the garden. But it's also a book of aspiration and inspiration, journeying through the many cycles of elemental ecology in a garden and how we can challenge our minds as much as our hands to think anew about the green spaces we tend. Without inspiration, there is no motivation. If gardening is only hard work, then the mind has ceased to believe in the cause. Bring the full force of the bitter winter winds and a gardener who knows that their endeavour is for the greater good will work long past the hour others give in.

The choices we make in our gardens today will be felt by many generations to come. Our gardens can become refuges for wildlife, and the lifeblood of a restitution of soil health and life. They can foster a rebalancing of our water systems towards a clean and more prosperous future, and they can aid in the regeneration of indigenous plant diversity and abundance. All will mean a great leap into the unknown where we live side by side with the natural world, and its provisions are not plundered nor its functions weakened. Gardens can become ecosystems like their cousins in the wild, moving and living dynamically and in harmony within the natural spaces they exist. Gardens are the portals to the regeneration of our bonds to the wild, and if we take the right path, their remedy will replenish us all.

CHAPTER 1

Design

There is no one single way to design a garden. Styles and traditions may have influence, but these are only motifs from which each garden inevitably departs. Every garden is unique, reflecting the environment and landscape it exists within as well as the hand of the gardener who dreamt of its creation and raised it from the ground. Gardens are dynamic; they grow, expand, reproduce, decompose, freeze, bake, dry out, and flood. Mammals, birds, insects, reptiles and microorganisms all live, breathe, sustain and perish within them. Not only do the seasons change but the climate is changing as well, and our gardens will transition with it, however extreme or unpredictable its effects. Gardens are, in short, ecosystems.

Rich, biodiverse environments are complex, but to design a garden ecologically is simple. It requires observing the space we currently garden, or plan to, studying its singular rhythm and unique patterns. It also means letting go of design ideas of the past by letting go of our desire to hold dominion over nature, allowing it instead to lead the way. We must recognise the cycles of upheaval and regeneration that are already ongoing in the garden and work with, not against, them. Nature is capable of producing infinite variety, but as gardeners and designers we hold grave power to diminish its divergency. With thoughtful vision and a light touch, however, we can rebalance the scales. If we design our gardens to be regenerative, the result will be functional, beautiful spaces full of life and vigour, robust enough to face the challenges of the future and elegant enough to beguile all those who walk among them.

Elemental Observation

Design begins in the eye and the mind long before it is sketched on paper or realised on the ground. To begin, it's important to understand the many and varied elements that make up a natural space. The ecological framework of a garden should be the foundation from which your design arises, and inform your gardening methods thereafter. You don't need to become a biologist or geologist, only a simple observer, gaining knowledge about your landscape as only you can, and imagining the possibilities from there. And nothing could be simpler than imagining your garden according to the four wild elements: earth, air, fire and water.

Knowing where the sun rises and sets will be vital even in the winter months.

I have found that garden designers often overlook these considerations in pursuit of aesthetical beauty. I too can be seduced by the power of form and craft, but I've learnt to train my eye to see beyond the surface to the foundations of garden construction. I've learnt this by seeing design not as patterns humans impose on

The Wild Elements

Observe your garden (or future garden) within the framework of the four elements by asking questions. The answers will then further guide your vision and design.

Earth

- What is the topography of the garden? Is the land on a slope or is it flat?
- What is the predominant vegetation of the local and wider environment? Is the land boggy and low-lying or is it by the sea? Does it sit within a temperate forest region or a dry steppe?
- What is the underlying subsoil? Is it heavy clay, or is it mainly sand or even chalk?
- What of the life within the topsoil – does it contain many worms and invertebrates, or is it lifeless and inert?

Fire

- From the perspective of the house wall, what direction does the garden face – south or north? Which area of the garden receives the most light?
- Is the garden overburdened by trees, high walls or other buildings, restricting its light? Will the lower winter sun reach over these obstacles?
- Do internal walls naturally collect and accumulate heat? And is the power of this heat energy detrimentally intense and needs calming, or relatively weak and needs amplifying?

Air

- What is your garden's prevailing localised climate? Is it located in a valley that holds morning fog? Does it experience high levels of humidity?
- Are there microclimates within the garden where frost pockets develop, and does dew collect in the mornings?
- What of the wider climatic zone – is the garden within a coastal region or a temperate forest?
- What direction does the prevailing wind come from? Does the garden require protection from regular strong wind, or should your design encourage wind to cool it down?

Water

- How much rainfall is the garden likely to receive – is water naturally abundant or do you treat it as precious?
- Where in the garden does water collect the most? Does it pool and linger or quickly drain away?
- Are there any watercourses that run through the garden? Any wells or springs?
- Does the water table come high in the winter or stay low the whole year through? Does the garden experience flood or drought, and if so, which areas are affected most by these extremes?

THE SOIL TEST

1. Dig a hole 50cm deep and take a handful of soil from the bottom. Make sure you dig below the topsoil level (figure 1).
2. Observe the soil. Is it sandy, or dense and heavy? Does it have rock in it or feel wet?
3. Take a thimble-sized amount, add a small amount of water or spit, and roll it into a ball (figure 2). If you can't do this, you have a sandy soil.
4. Take the ball, roll it into a sausage shape and loop this into a circle (figure 3). If it breaks apart, then you have loam.
5. If you have been able to form a loop with no cracks in it, then your soil is clay.

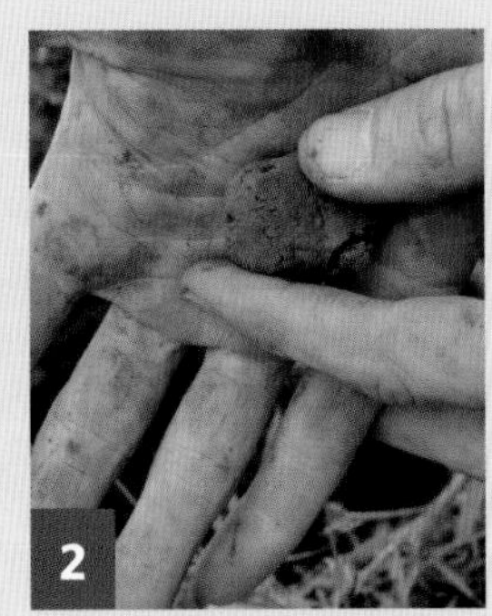

nature, but the fabric of how nature itself works. Every design I have ever drawn has been improved the longer I've spent getting to know the space the garden will occupy. In your own space, gaining knowledge of certain conditions in your garden may require slightly more than passive observation. For example, to better understand your underlying subsoil, you can conduct two simple soil and water tests outlined in The Soil Test and The Water Test.

THE WATER TEST

1. Dig a hole 50cm deep and a spade's width wide. Make sure you choose a site on level ground.
2. When no rain is forecast, fill the hole to the top with harvested rainwater and check regularly (figure 1).
3. If the water drains away quickly, the soil is loose and free-draining. But if it lingers for a long time, the soil is heavy and saturated.

1

Visualising the Future

While initial observations can be great for gathering insight and wisdom, the ability to visualise how a garden will act, look and grow in the future is what will set a design apart. A garden is a living entity, always growing, changing and adapting through the seasons, which in turn are driven by the local climate. It's impossible to know exactly what will happen to the climate in the future, but we can all assume that the steady state of Earth's system can no longer be relied upon. Extreme weather events will be more frequent and only the most robust and adaptable gardens will flourish. The goal should be to reverse these trends, but we can also create resilient gardens and give more to the generations that follow us, rather than leave them with less than we originally acquired. A tree, for example, can grow and provide for hundreds of years, but when choosing what species to plant and where, there are a number of important considerations. How fast will the tree grow? How much shade will it generate? How much water will it take out of the soil? And will it increase or decrease the garden's biodiversity as it ages? All the answers matter to the design and all must be considered from the moment the sapling goes into the ground.

Imagine, for example, that an area of your garden was both prone to winter flooding and situated near a pond. In this case, you would want to plant trees to add structure to the soil and soak up the excess water, but you wouldn't want them to be so tall as to overshadow the pond. Dogwoods (*Cornus* spp.) and willows (*Salix* spp.) both thrive in saturated ground and absorb much of the water, allowing for a wider variety of species to colonise the space around them. They coppice and pollard extremely well too, so you can restrict their height so as to not overshadow the pond, further expanding its biodiversity.

When designing for the future, we must consider not only trees but also the rapid growers and over-abundant colonisers. If we want to encourage a certain species to expand and colonise an area over many years, like daffodils, for example, then we must allow space in the design for this to happen naturally. Plants are opportunistic and will always find ways to defy the limits of our designs, but we can avoid unwanted outcomes in the future – like plants spreading where we don't intend them to – with better forethought at the outset.

We must also envision how a garden will look in the drastically different guises of winter and summer; I have always found myself as surprised by the expansive growth of summer as by the

The garden can be a radically different place in winter.

skeletal emptiness of winter. Seasonal changes are natural, but it's easy to forget how starkly altered a garden can look; if we design our garden to accommodate the rapid expansion of summer growth, we must also imagine how far the growth retreats. Winter leaves only the bones, or structure, of the garden – essentially the outline of the design. It's important to contemplate and retain the state of your garden in winter, not only for aesthetic reasons but also for the benefit of overwintering or winter-visiting wildlife – such as keeping seed heads on perennials and leaving hedgerows uncut until spring. Since nothing ever stops in a garden, including ourselves, and winters are long and drawn-out depending on latitude, the garden still needs to function through the winter and also hold some charm and beauty to enliven both the soul and mind of the gardener who works within it.

Drawings

At some point, however, thoughts and visions need committing to paper before time diminishes their power. It's only once plans are drawn that a cohesive and holistic design can be achieved throughout the garden and this must be the ultimate aim. Begin by creating an outline drawing, as described in Creating Outline Drawings.

Once you have the basic outline drawing, you can now play and experiment with your vision for the space, but keep in mind that drawing is the place for ideas, not artistic endeavour. It will also never substitute the vision that can be gained from standing in the garden space, and the ideal scenario for the ecological gardener is to use both in conjunction to reach the most desired outcome.

I cannot emphasise enough how few further skills of garden design drawing are needed beyond this point. If you have the outline and are working in scale, being loose and free will be your greatest asset. Do it your own way in your own style and don't follow any classical ideas of how it should or shouldn't be drawn. As part of my profession, I have seen some of the worst gardens result from the most immaculate drawings, and some of the most exceptional gardens emerge from the childlike scribbles of the

CREATING OUTLINE DRAWINGS

Outline drawings rely on accurate measurements and using reel tape is a cheap and effective way of achieving them. However, if you have a particularly large garden with varied topography, then using a rotary laser level can be of great advantage, though a costly one.

1. Measure the perimeter of the garden with a 30-50m reel tape. Secure the tape at one end, laying it flat on the ground (figure 1). Be sure to measure everything twice before noting it down (figure 2).
2. Roughly sketch the perimeter in a notebook, logging the lengths along each boundary line.
3. Measure and mark any permanent internal structures of the garden - trees, greenhouses, etc.
4. Transfer the recorded measurements and shape onto A3 paper within a scale - usually 50:1 or 100:1 (you will need a scaled ruler for this). Transfer measurements in pencil first, then with pen (figure 3).

1

2

3

best designers. This doesn't mean that drawing is irrelevant – far from it. I only mean to blow away popular misconceptions that it takes great skill to undertake drawings. We are trying to create gardens, not architecture, and the inspiration can be as much use as the actual drawing. Gather magazine cuttings or create boards, visit parks or gardens that you admire, and read books and watch videos to help distil your vision into reality. You will always want more for your garden than space allows, unless you are very fortunate, so this process will become more about selective restraint than ample gluttony.

Cyclical Design

The natural world's rhythms are governed by the cycles of the year, and a garden is no different. In the temperate UK, seasons are pronounced and contrasting, allowing for vast differences in heat, light and rainfall. Cyclical design acknowledges these changes and looks to work within, not against, their assumed regularity. Even though a blank canvas offers us the opportunity for personal expression and artistic liberty, our designs must accordingly put ecological considerations above all else. This is crucial to understand and abide by, as the balance has forever been tilted in the other direction, towards aesthetics, with ecology secondary if even considered at all. Although the aim is not to be dogmatic, ecology should be the foundation on which the rest of your garden design decisions will depend. You can begin by identifying what will comprise the heart of your garden in the form of primary components (ponds, greenhouses, etc.). Once identified, the connections needed to link these components, including paths and swales, will become apparent and a cyclical order will start to take hold.

Main Components

Greenhouses, ponds and compost piles are all used by the gardener to enable the functions that make the rest of the garden flourish. Once you've decided which of these to include in your design, place them in areas of the garden that will depend on them the most, where they will form the most beneficial relationships. For example, you might situate a pond next to an orchard or a coppice, using the pond to water them in the summer months. Or you might plan to keep your compost area adjacent to the vegetable garden, where its use is most frequent.

As they do in nature, such partnerships – some obvious, some subtle – can happen all over the garden. The more time you spend in the garden, and recurring tasks unfold, the more opportunities for thoughtful, holistic design will manifest themselves. The engine of an ecological garden's success and vitality will always be about production, whether that be creating compost for the soil or water for the plants. By recognising this, we can design for the future health and robustness of the garden, prioritising its working functions over aesthetical rewards.

Connections

Just as important as the main features of the garden are the paths, rills, gullies and swales used to connect them together. Paths are far more than simply functional; they define the journey through the garden, form its axis, and can themselves be a focal point. They can be permanent, like a brick path leading from a door, or transient, like a path mown through a meadow that changes course from one year to the next. When deciding on materials, it's important to consider how frequently a path will be used; a path to a coppice at the edge of the garden might receive little traffic, while a path between a compost pile and a border would see regular use even in the worst of weathers. But however and wherever they are placed, paths will create delineation, line and shape within the garden space, and the rest of the garden's design will conform to them. The shortest distance between A and B is not always the most desirable, so try to follow contours and weave around naturally occurring elements like trees or depressions. All the best roads are forced to bend to nature's will and paths are no different. The rills, gullies and swales we can use to carry water through the garden will have the same effect of delineation and line, and they too will benefit from a delayed course as they ribbon from one body of water to the next.

I have found that people all too often seek to oversimplify when solving problems, and designers are no different. After all, what can be more efficient than planning pathways using only straight lines, right angles and curves? And yet some of the most pleasing elements of any garden are those that juxtapose the influences of nature and the human hand. It's far too easy to landscape away the unique contours of the ground that have taken hundreds if not thousands of years to form, yet I know from experience that doing so results in the loss of a natural weathered charm that is missed when its importance is overlooked.

A rill running the overflow out from a pond.

Prevailing Climate

While it's important to design your garden so that its main features work in balance with one another (e.g. compost areas near vegetable gardens), possibly of greater importance is placing certain areas according to your garden's prevailing climate. For instance, greenhouses should be located away from all shade, and face south to receive the maximum amount of light and heat energy from the sun. All over every garden, microclimates can be harnessed and utilised, minimising the amount of additional input we need to give. In the main, designing for climate is about amplifying natural forces, but it can also be about limiting their power or even creating new microclimates ourselves. Common sense will guide many of our decisions in this regard, but it isn't always easy to glean the right amount of light and energy for their obvious benefits without risking the negative effects of too much heat. Getting the right balance, however, is of vital importance to form gardens that will, over time, become more self-sustaining and freely regulating. The key to this is to use the knowledge gained from your observation and study of the site to guide your design.

Wildflower meadows, for example, require as much light as possible, as do orchards and coppices, which find unencumbered exposure to the elements most salubrious. Other areas in your garden, such as seating areas, will benefit more from a filtering of the sun's rays, with a pergola and climbing vine providing light shade. Most plantings too will flourish away from constant sunlight, working best off east or west aspects or under small deciduous trees. Of course, many plants will thrive in the lowest of light situations, but I wouldn't be inclined to design towards that fact, instead utilising these adaptable characteristics when the situation arises.

It isn't just sunlight that needs to be considered for the wider ecology of the garden; wind too needs to be blocked, diverted and mitigated against. Wind increases evaporation on leaves in dry conditions and can also lower temperatures in the colder months. Windbreaks in the form of trees and hedges will probably do more

1

MARKING OUT A PATH

To create paths in the garden that work in balance and harmony with the natural contours of the land, it's important to mark out and visualise how the path will need to adapt before proceeding.

1. Consider the type of path you want to create, taking inspiration from examples you happen to find elsewhere. In this example, we are making a wide natural grass path through a grove of trees (figure 1).
2. Visualise the length and width of the path based on the space available and the eventual height and spread of the trees.
3. Place canes into the ground running the desired length of the path and repeat along both edges. Use large canes and make sure they are secured in the ground (figure 2).
4. Bind twine between each cane to create an extra visualisation of the line the trees will be planted along. Be sure to cut the twine at the length of the entire path (figure 3).
5. Adjust the canes until they represent the line and shape you desire (figure 4).

than anything to create beneficial microclimates within the garden, but don't attempt to enclose your whole space, as wind acts as an antifungal in trees and plants. It is also the only way you can create movement in the garden, other than with water, so defend against its elemental force with an even hand and allow some areas to be open and others enclosed.

No ecological design can be truly effective without water at its heart. We will cover in great detail the methods and means of collecting and reusing water later in the book, but good design should involve finding the most advantageous place for water to be. Ponds, for example, are best placed where natural depressions or pools of water collect, while swales should be sited on the contour of slight slopes to capture the rain running off the surface. The eventual aim of water design is to use gravity to journey the water through the garden as slowly as possible, with ponds linked by rills, gullies, pipes and swales. A pond can even create its own microclimate by reflecting sunlight and increasing humidity in the summer, using its energy in a secondary form to further benefit the garden environment.

Every gardener forms an intimate bond with the weather and prevailing climate of their region, but it still surprises me how many fight against their conditions rather than work with them. I have found myself equally perplexed and dumbfounded to be standing in a greenhouse in the depths of winter with a temperature heater roaring, or in the same greenhouse, devoid of plants, at the height of summer. I understand the need to adjust to the extremes that the climate can bring, but it's by careful planning of storing heat through the winter and cooling it down in the summer that will enable continued good practice.

Closed-Loop Design

In addition to considering the elements, the goal of ecological design is a garden that is dependent only on its own resources. Waste has not been seen as a complete resource, and while some of this is down to institutionalised thinking, much is also the result of poor design. Simplistically a no-waste system is one where all the outlay from one function becomes the resource of another. While no-waste systems can be difficult to achieve in other areas of our lives, there is no greater place to put the no-waste mantra into action than in our gardens. I have spent many years as a professional gardener watching on as wood is burnt in huge mounds, water is swept away down drains, and stones and bricks discarded in skips. Such wastefulness is common, but can be easily bypassed by implementing closed-loop, no-waste designs.

The powerhouse of the no-waste system is the compost area. The key here is to estimate the volumes of organic material your garden will absorb and design with this in mind. The compost heap will need to take on everything and anything that can be composted from the house and garden – an amount that is far more than people ever imagine. It must also be periodically turned to raise its internal temperature, so we must imagine, and design, enough space for many heaps – possibly adjoined in bays. This entire area can then include vermicomposting houses, biochar burning pits and other subsidiary composting designs that can feed into the main compost heap.

Water too is a major cyclical resource; we must make the most of all the water we collect by feeding it through as many ponds, swales and rain gardens as possible before it finally leaves the garden and percolates back into underground aquifers. More than this, however, it is important to design a system that considers the grey water from the house, as this will require a separate pathway through a series of filtering stages or collections. At every stage, water can grow more plants, create more life, and restock the underground water tables.

In addition, when putting your designs into practice, plan to work with materials that you already find naturally occurring in the garden (wood, clay, stone, etc.), or are reclaimed from other sources, rather than plastic or store-bought. Crucially, garden materials should also have the ability either to be reused indefinitely or to see out their lifetimes in the garden and biodegrade into the soils. A no-waste design facilitates the action and method

Designing Interdependency: Chickens

Raising chickens, whether for meat or eggs (or both), offers a clear example of interdependent design, where each stage of the cycle results in an output that prompts the next.

Stage 1. Food scraps are fed to chickens in pen/coop → Nutrient-rich manure compost
Stage 2. Compost is added to the soil → Fresh organic produce
Stage 3. Organic produce is cut and consumed → Food scraps for the chickens

needed for a no-waste system to work, but in my experience, the ecological gardener has the greatest chance of success if both are working in unison, allowing the system to run.

Nature's Blueprint

When asked to think of the most abundant ecosystem on Earth, most people will imagine a tropical rainforest and its staggering amount of biodiversity. Approximating such abundance requires designing your garden to allow – and/or aid in producing – more sunlight, more water and more organic matter. This, in turn, will produce more plant growth, draw more carbon from the air, foster more wildlife and microorganisms, and result in more biodiversity. If this seems complex or unattainable, it's actually the complete opposite; it's the simplest way to design and shepherd a garden.

To attain a garden full of biodiversity and abundance, the design should abandon minimalism and monocultures in favour of complex species-rich environments. Lawns can be regenerated into wildflower meadows; mono-species hedges turned into mixed-species hedgerows; ponds left to develop into self-sustaining wildlife habitats; and native plants encouraged over exotic cultivars. Rewilding a garden is as much about letting go in your design as it is about assisting and abetting. Once you have harnessed the elements and put the apparatuses in place to aid nature's working functions, you can loosen the reins and be free to play with the most compelling of all the garden's ingredients: the plants.

In the natural world, plants compete with one another for light, nutrients and other resources, and in many ways, we have lacked the forethought to account for this in our planting designs. Traditionally, planting plans are drawn with an idea that individual plants will grow bigger but that their place is fixed. When interloping species or rampant spreaders and seeders diverge from the design, they therefore need to be checked or removed. But plants are dynamic and opportunistic, and if we curtail their instincts completely, we cut through their DNA and leave them adrift from their natural habitat. Therefore our designs must be dynamic too, allowing for more flexibility and fluidity to aid the plants' natural urge to reproduce and expand or pioneer new ground. This will benefit the ecosystem of the garden in more varied ways and lead to more richly biodiverse plantings.

Successional Design

Imagine a lawn. Now imagine that same lawn with a tree growing in it, a vine climbing through its branches, herbs and perennials reaching up its trunk, and a carpet of bulbs and creepers covering its feet. Which one is more biodiverse?

And yet if all these plants were to reach their apex at the same time, they would equally suffer from the unnatural competition this would create. However, if we design so that each plant reaches its apex at different times of the year, they will grow with

Designing in Layers

Designing with vertical space in mind, not just horizontal, allows you to achieve more diversity in your planting in the smallest amount of space.

Lower layer: Prostrate plants that expand over the ground, barely reaching above ankle height, such as ground ivy, ramsons (figure 1) and primroses.

Middle layer: Upright and bushy plants from a vast array of groups – bulbs, annuals, biennials, perennials, herbs, grasses and shrubs – reaching from ankle to head height. For example, foxgloves (a biennial that grows from seed set in the previous year, figure 2), quaking grass and common broom.

Upper layer: Plants that reach above head height but are not trees, usually confined to shrubs, grasses and small climbers such as dogwood, sweet briar, guelder rose (a magnet to wildlife and an ancient-woodland indicator, figure 3) and common reed.

Canopy layer: Trees, with the occasional vertical climber using the trees to reach the canopy. In a garden setting, think of trees such as field maple and hazel, with climbers such as dog rose (figure 4), hops and ivy.

more vigour and vitality. If the bulbs come into flower in the spring before the tree is in leaf and the perennials grow too high, they will have access to more light and energy, which they can use to grow healthily, reproduce and expand more readily. Their nectar, too, will be available at a time when it is otherwise scarce and the birds and bees need it the most. Then as the bulbs go over and the tree expands to full leaf, the perennials – which are suited to dappled shade – will grow tall and produce their leaves and flowers, offering sanctuary and sustenance. As the season changes, the tree will let more light in as it loses its leaves, and new sets of bulbs and perennials take the place of others that have since faded, prolonging the diversity of the area across the whole growing season. Even as the growing ceases and the winter takes hold, if we design our plans with plants that will hold their form and structure even in decay, then we will provide further sanctuary to the wildlife, maintaining biodiversity.

Although this type of design is not always correct or desirable for every scenario, making the most of the space available to create abundance is a key principle. If your garden features areas of hard standing like terraces and paths, place pots at their edges and plant them up in a miniature layer of succession. You can also plant species suited to dry, free-draining environments along the sides and tops of walls, and plant succulents on flat roofs, expanding the ecosystem beyond the garden floor. Also, if you don't have a tree or wall for climbers and vines to scramble up, make upright supports or arches from coppiced hazel, willow or reclaimed metals. Gardens are full of more possibilities than at first it would seem, a fact that may be easier to realise if you draw your design on paper first. It may be more useful to draw vertical designs in perspective 3D rather than outline 2D, but if that's not possible, then just keep visualising your ideas on the ground, marking out height differences with canes and poles. Designing with vertical ascent in mind will enhance the depth of any space as well as the diversity and abundance.

Most garden owners I work with support this ideology, understanding the need for plantings to be rich and lush. But almost all stop short of the rampant anarchy that is actually needed. This conservatism in the garden may be due to the difficulty of imagining randomness and complexity, but it is also rooted in traditional beliefs about the wisdom of control and, to a lesser extent, minimalism. I cannot always persuade garden owners against the wishes and desires they have for their own spaces, but as awareness grows for the role that layered abundance plays in the restoration of ecology, the more this method will be embraced and applied.

Wild DNA

If we want our plantings to thrive and be healthy, we must observe where they dwell in the natural world and place them in a similar context in the garden. Say, for example, there is a low-lying area in your garden near a watercourse like a river or a stream. What plants exist in a similar scenario in the natural world? Those will be the ones you'll want to use. In doing so, not only will you recreate the ideal conditions for those plants to flourish, but you'll also provide a small patch of habitat for riparian wildlife, like newts and frogs, increasing the likelihood of their presence and aiding their recovery from continued decline in the wild. This lies at the heart of a regenerative ethos.

In addition to observing and recreating plants' natural habitats, we can also learn from the relationships they form with other plants in those habitats. For example, hedgerows feature a wide tapestry of species that have evolved together in harmony over hundreds if not thousands of years. That they have existed alongside each other for so long means that within a similar environment they are unlikely to dominate one another, and will transition to holistic harmony and balance more readily. Some combinations of species will even hold beneficial relationships far more complex than what we can observe, with their ingrained, evolutionary interdependence facilitating each other's survival. Whether underground via the complex matrix of symbiotic and mutually dependent fungal pathways, or through defences against predators and pests that some species avail to others, mutualistic

Ferns and assorted other plants thriving in the dappled shade.

communities, by their example, offer us a pathway to follow for our gardens; and if we want to build similar ecosystems into our own designs, then we must understand these relationships.

Plants' reproduction and expansion habits can also inform our considerations and designs. If we notice a species, like nettles (*Urtica dioica*), has taken over a space in a dense mass, it's likely to have done so through natural expansion. However, if its habit is to drift across an area loosely and randomly, like dandelions (*Taraxacum officinale*), then it is likely to have done so by seed. When incorporating indigenous native species in our designs, it's important to understand plants in this way rather than to superficially observe their 'look' in situ. This is because many garden varieties of plants have had these characteristics extracted or subdued through cultivation, precisely because breeders want these plants to stay put and not enact their natural tendencies. We must be brave, however, and give native plants the space to reproduce naturally.

I have only come to this understanding from endless trial and error, designing on the fly and accepting that even though some structures and features of the garden are more permanent and fixed, plants need to transition, move and be free. Wandering wildflowers are opportunists and I have always admired them for this tenacity and endurance. It took developing a complete disregard for the traditions of gardening craft for me to stop trying to abate this troubadour spirit and instead follow their lead, letting the garden be fluid and open. In other words, designing to rewild, to unchain, and to abandon neatness and order. Over the years, I have noticed that neatness has become a default for achievement in gardening, as though presenting the garden in a tidy fashion makes it somehow more illustrious and grand. But dominion over nature is fool's gold, with the cost far higher than the perceived reward.

Ode to Joy: Designing for Ourselves

Once all ecological considerations have been met and your garden's design mirrors the form and function of a living ecosystem, you can then turn your attention to more aesthetical concerns. As gardeners, as long as our desires and wishes don't have a detrimental effect on the system we have created, then we will fulfil our role as the keystone creator and sculptor of the garden. After all, it is only through our wonder at the beauty of nature that we come to admire and revere it, and then look to cultivate its continued health and prosperity. And yet the aesthetics of the human hand can be equally admired in the design and composition of garden space, with line, form and structure creating their own pleasing harmony. Aesthetics can also be beneficial for our minds and consciousness, alleviating and nourishing in equal measure through our primal connection with natural space, and for our bodies, through the physical activity of gardening as well as from the sustenance a garden can provide.

For the Eye

Good design, on a purely visual level, means many things to many different people. What you, and you alone, find appealing and sublime should be the guide you follow. However, there are some ideals that have stood the test of time and have become universal in their ability to please the human eye. Chief among these are vistas or views; at any point within the garden a harmonious view can unfold like a painting on a canvas, whether it's out across a pond to the landscape beyond or down an avenue of trees to a perfectly placed pot. Vistas usually follow along an axis line that runs through the garden from one end to the other, taking their starting point from doors, windows or the edges of the house. In addition to views, the line, shape and overall form of a garden are also important for aesthetic beauty. You can enhance paths beyond their utilitarian function by adding detailed edging or charting their course along axis lines. Similarly, you can design the rills and gullies that carry water through your garden to run parallel to paths, walls or avenues. At the edges of terraces, paths and walls, borders can be placed to further emphasise the outline, while outer boundaries can be delineated by permeable fences.

This view reaches through the pergola to the stone lintel and beyond.

Tips for Creating a View or Vista

1. Open up a section of a dense boundary so that you can see out beyond.
2. Stagger or partition hedges instead of creating continuous blocks.
3. Run paths, steps, gates and doors along the same axis lines.
4. Position tall focal points such as trees at the opposing ends of wide, level spaces like ponds and meadows.

Next to consider are colour, movement and structure. Colour belongs not only to the flowers and the foliage, but also to the materials you use in the garden and the landscape. However, since plants tend to dominate our attention when designing a garden, it is paramount to remember that most plants flower only for a matter of weeks but will be in leaf for months or perpetually, so when designing around colour, consider the foliage of a plant first. Whether it be foliage or flower, the key is to remember harmony and balance. Harsh contrasting colours can work occasionally but, in the main, colours that complement each other will win out in the end. Think like a painter, who uses a restricted palette in a vast array of hues and shades to bring an artwork together; a garden should work in the same way. The shape and texture of plants, on the other hand, should be contrasted as much as possible, as too many leaves and flowers of the same shape can look contrived and unnatural. Try combining upright plants, like purple loosestrife (*Lythrum salicaria*), with those that have arching form, such as the lady fern (*Athyrium filix-femina*); or pairing big rounded flowers with wispy, ephemeral ones. Using materials with contrasting shapes and textures – like linking a drystone wall with a reclaimed wooden step, or erecting a reclaimed metal arbour over a flagstone seat – will also enhance the design. And, of course, plants will contrast with materials in both texture and form in an infinite number of ways.

The result will be a natural space that has been adapted by the human hand, and although ecological gardening is at heart about designing under the guise of nature's wing, the 'bones' of the garden will always come through. Every outline, shape and delineation will be amplified, especially when the growing season fades and winter takes hold, so if you design to anticipate how the garden will look throughout the whole year, it will be better for it.

For the Mind

As much as a garden can offer great visual delights, its gifts can be even more profound and rich in the realm of our mind. It's no surprise that gardens and gardening are seen as places of mental health restitution and many wonderful gardens have been built in recent years with this as their sole purpose. Gardens soothe the edges of our mind and can uplift, nourish and enlighten those of us with the steadiest of mental constitutions. Our designs can encompass elements such as privacy and sanctuary from the chaos of a frenetic world, and offer the space and openness for mindfulness and expansive thought. They can also be places of reflection and calm tranquillity, especially those that feature ponds or running water, and foster a sense of wonder and connection with nature. Gardening itself can also be a cathartic cerebral endeavour.

When I am alone working in the garden, I am totally and completely absorbed by the task at hand. None of the digressions of life that filled my mind earlier in the day linger and a certain sense of calm prevails. I don't experience total relaxation, as I am working towards completing a task, but rather a steadier state of peaceful activity. This feeling is always augmented and heightened if the garden has a sense of privacy and space from the outside world, and it's the subtle art of design that allows this to happen.

To allow a garden to have both space and privacy, follow a simple mantra of dimensions: horizontals for space, verticals for privacy. More horizontal features – including meadows, terraces, paths, ponds and clearings – will stretch space along the ground, opening the garden to the sky and views of the surrounding landscape. They allow more light and air into the garden, making it feel more connected with the outside world and the changing patterns of weather. When you employ a lot of horizontal elements in your design, the garden will feel bigger as a result and lend itself well to contemplation, much like a seascape or great plain. Historically, this expansiveness has been achieved through the use of lawns, but lawns are anathema to biodiversity so meadows are far superior for this purpose. If you want to encourage a more meditative experience in the garden, you can create permanent areas, thoughtfully placed, for people to sit and take in the view. Seating is best when it is vernacular in style and blends seamlessly with its surroundings, such as a wooden chair made from a fallen tree or a low bench crafted from the rubble stone found in the ground.

For more privacy, the key is to think of vertical elements less as impregnable barriers and more like visual tricks to make the garden feel more intimate but without the impenetrability that blocks out wildlife. These can include hedges, walls, pergolas, trees and doors. Too often privacy is used as a means to enclose ourselves from invasion, but this should not be the aim; the only thing we need is a place away from prying eyes where we can feel safe and turn down the dial of the outside world. Hedges can work well along boundaries for privacy but they can also be wonderfully effective internally, increasing its sense of secrecy and mystery. Fences, too, can serve the same function, but remember they should be spaced and open, like a trellis with climbing plants rambling all over it. Trees can encircle a space, especially along boundaries, but smaller trees can also work well within the garden to evoke a feeling of private seclusion. Walls are more solid constructions that will completely envelop a space and increase its seclusion, and also sometimes aid in creating microclimates through their ability to retain heat. Doors are wonderful in walls, hedges or fences and can add to the sense of mystery and sanctuary that being closed in the garden behind them can bring.

A yew hedge can grow to an imposing size, offering total privacy.

For the Body

Finally, gardens can be designed for the effect they have on our bodies as much as on our minds and eyes. For one, gardening requires physical activity, and different designs can encourage different levels of participation and exertion, contributing to the gardener's health and well-being. For example, a garden stocked with fruit trees, country hedges and climbing vines will need more pruning in the winter months than a garden of more simple plantings, involving extra physical effort as a consequence.

In my opinion, the greatest gift a garden can give is that it regenerates us as much as we regenerate it. Gardens also offer sustenance to our bodies to sustain this relationship. Fruits and vegetables that can nourish us and our families can be produced in a garden as easily as any farmer's field. You can dedicate areas such as veg gardens and food forests to this, but you can also weave food crops in with your flowers to make a seamless tapestry. Either way, a garden of sustenance is good for the soul and can provide energy for an active life. And if we can grow it in the garden, we can eat it there too. Dining areas can be the most convivial of all the garden areas, as they are places we can share the harvest with others. Sitting at a table and eating great food on a long summer's evening under a pergola bedecked in scented climbers is joyful. We can design our terrace areas with such a scene in mind, even if it only comes to pass once in a while.

CHAPTER 2

Soil

Dirt, earth, ground, mud, land, muck, mire, terra firma, crud. If ever there was a friend to humans, it would be soil. In the jungles and the forests, under the grasslands and in the bogs and fens, soil is the driver. Without soil the land would be a barren salt pan, an inert desert, or an ice-sheathed mountain. It's rudimentary to the existence of abundant life, and yet it seems we are still just beginning to unravel its mysteries and understand its delicate balance. As gardeners we would say we know it only too well, but do we really know what lies beneath our feet? Most of us have cracked a spade through its surface and planted within its earthen basket, but how much do we know of the microorganisms? Of the symbiotic lines of communication that fungi thread from root to root? And what about the invisible powers of minerals and gases under our feet? We are the guardians of our soil; in order to protect and guide it towards regeneration and nourishment, we must see it for the vital resource and foundation of the garden ecosystem that it is.

Soils are ecosystems in their own right: the media and conduits that all the elements of their composition orbit through. In

There are more microorganisms in one handful of soil than there are people on Earth.

their alluvial alchemy live a vast rainbow of microorganisms, including bacteria, fungi, protozoa, archaea and algae, the engines of a factory of death and rebirth, decay and renewal. These organisms are the lifeblood of the soil, their actions providing nutrients for plants to grow and, in turn, harbour sanctuary and sustenance to a dazzling array of life – including ourselves. The balance of our atmosphere is also maintained in part by their endless endeavour, as they return some of the carbon dioxide that plants draw down to store and further aid their growth. Mycorrhizae (root fungi), for example, extend the reach of plants' root systems to give them more nutrients and water in return for some of the plants' carbohydrates. These fungi have hair-like filaments, or roots, that expand out so far as to become connected to the surrounding plants, and can divert sustenance to ailing or younger plants, or even warn of impending dangers – an extraordinary web of communication and collective design. Soil biologists, especially those who study at the microorganism level, are still on the frontiers of discovery and knowledge; soil's complex patterns of existence are only just beginning to reveal to us their long-held secrets and wonders.

It's all too easy as gardeners to forget about the microscopic life in the soil as we go about our tasks. I have, in the past, certainly been guilty of seeing the aeration of soils as a good thing, breaking up lumpy masses into workable friable chunks. How wrong I was. Once I learnt more about the no-dig methods and

Fungi, along with bacteria, are the great decomposers of organic matter.

theory, and microorganism activity became more well known, I admonished myself for the obviousness of it all. In the wild, digging is rare and digging to depth rarer still. I now understand that leaving the soil untouched as much as possible will reap far greater benefits than any amount of aeration, and that allowing soils to weave their complex matrixes is more important than anything else. Sometimes we must reject what we have been taught to open our thoughts to new discoveries and progressive ideas and information.

Soil, as a structure, holds both the roots of the tiniest seedling and the colossal digits of the mightiest tree; almost every plant owes its survival to soil's super-power to support any mass within its cradle. Soil is also a repository of such fundamental elements of life as carbon, nitrogen, phosphorous, magnesium, calcium, iron, zinc and boron – all essential to plant health and growth, and to humans and other animals too. The majority of our freshwater also flows through its remarkable framework. In our gardens, rainwater percolates down through the soil into vital underground aquifers and reservoirs, which in turn feed the rivers, streams and estuaries. Lastly, soil is home to a myriad of invertebrate and insect life – worms, spiders, beetles, bees, millipedes, centipedes, snails and many, many more. These, in turn, are a food source for birds, amphibians, reptiles and mammals.

Armed with this knowledge, what can we do as conscientious gardeners to shepherd our soils into a balanced state? Firstly, it should be self-evident that bare soil – or worse still, tilled bare soil – left open to the elements is an entirely unnatural phenomenon. It's estimated the Earth's soils hold 2,500 gigatons of carbon – over three times the amount of carbon in the atmosphere and four times the amount stored in all living plants and animals. Bare soil emits far higher amounts of carbon dioxide to the atmosphere, and supports fewer microorganisms. Covering our soils with mulches and avoiding unnecessary or excessive digging will reverse the damages wrought by digging by more closely replicating the conditions seen in the wild. In addition, avoid activity that will lead to compaction, like tracking heavy machinery or continued treading; looser soils can absorb more water and provide more air for plant roots and passage for wildlife. These practices also help avoid erosion by slowing down the water that falls and flows through the garden.

Soil is the basket that holds the roots of the mightiest organisms in the world: trees.

In the natural world, the top three centimetres of soil can take up to a thousand years to grow, and it's worth keeping in mind how unique and fundamental the soil is in our humble plots. The degradation of that top layer of soil is continuing at a frightening pace and will have untold consequences, but by tending our soil in an organic, regenerative and ecologically minded way, we can create more holistic gardens in tune with their place in the natural world.

Composting Methods

Nothing opens the door to the complex mysteries of life and death more than understanding how compost works and guiding the process with your own hands. In a woodland, the forest floor absorbs everything that dies, turning it into soil, which then gives sustenance to what is living. Composting is about emulating that very process. While cycles of decay and renewal play out on their own, composting speeds up the process from what can sometimes take many decades to mere weeks or months.

When examining in great detail the method and practice of creating compost, it is also vitally important we don't forget to ask why. If a garden doesn't have a compost heap, then it doesn't really exist as a garden at all. It could be left completely to the wild, and that has some obvious benefits, but is that a garden either? A garden is defined by the interaction of the human hand with the natural world, not without, and certainly not against. If we are integrated within the ecological system, then we must play our part in perfect harmony with our environment and composting does exactly that. If you don't have a working, effective compost pile, then at some point you will need to buy some in, and at that moment you will be stepping outside the closed-loop system.

Compost, and therefore soil, is the heart and soul of the garden and its proper management a core principle of gardening organically and in harmony with nature. It's also deviously complex for its outwardly simple appearance. To say we understand everything about soil or compost would be like saying we fully understand the craters of the Moon or depths of the Mariana Trench. We don't. Every day we are learning more about soil's infinite variety and microscopic marvels, and as gardeners we must respect that fact

Making compost is to mimic the organic process of nature.

Table 2.1. Common Carbon and Nitrogen Compost Ingredients

Carbon	Nitrogen
Cardboard	Coffee grounds
Dry leaves	Fruit peel
Newspaper	Grass clippings
Soil	Pond weed
Twigs	Soft plants
Wood ash	Veg scraps

and be open to abandoning everything we thought we knew for what we are only just beginning to uncover. Compost is a living, breathing, organic, natural machine that can feed the soil and sustain our world. All we need to do to start is follow the recipe.

At the risk of oversimplifying, anything organic or derived from an organic source can be composted, be that leather gloves, fish heads, toenails, latex condoms or dead flies. Of course, composting these items is not the place to start, but it's worth noting that everything we throw away is someone else's problem and the better you get at composting the more you'll be able to put into the compost heap. All the materials you will most likely compost can be considered either one of just two ingredients: those that supply carbon (aka brown) and those that supply nitrogen (aka green) to the pile. Table 2.1 lists the most common forms of each that you will come across. When building a pile, aim for a fairly basic 50/50 ratio between the nitrogen and carbon ingredients. From here, begin to 'make' the compost.

Compost often suffers from a bad case of the PR blues, and an over-abundance of sterile scientific terms doesn't exactly help its cause. Yet in the absence of more holistic classification, some scientific terms are essential for understanding the different types of composting methods and how they work.

There are two main methods of composting: aerobic (with oxygen) and anaerobic (without oxygen). When done correctly, both – while distinctly unique – lead to the same outcome of nutrient-rich compost for your garden. Lest we forget that by composting we are only mimicking processes found in nature, think of aerobic composting as what happens on the forest floor, which is open to the elements of rain, air, heat, and microorganism and animal activity. Anaerobic composting corresponds instead to

A forest floor is the perfect example of aerobic composting.

a bog or sediment layer, where decaying matter is trapped beneath the water or surface and starved of air and oxygen.

You might also sometimes hear compost referred to as 'hot' or 'cold', so let's clarify these terms a little. Both refer to aerobic, rather than anaerobic, composting methods. 'Cold' aerobic composting is when you leave a pile of materials untouched, either in bins or on the ground, to mature into nutrient-rich compost. This process takes a long time – between one and two years. While cold composting requires less labour, because the pile's temperature is low, it will not be sufficient to kill all weed seeds and pathogens – a potential problem when you go to use the compost on your beds. 'Hot' aerobic composting entails heating the pile to a certain weed- and pathogen-killing temperature through the art of correct guidance, whether by turning and aerating the pile periodically, or the specific ingredients and their ratios, or both. Basically, 'hot' and 'cold' can be thought of as active and passive maintenance, so in this book when I refer to aerobic composting, it will always be the 'hot' variety.

The actual composting of your pile of leaves, scraps, clippings and other materials is done by microorganisms, mainly bacteria and fungi; your job is to assist them in their functions. Their primary requirements are water, oxygen and heat. Water is provided by the nitrogen-rich ingredients and weather, while oxygen can be increased by adding carbon materials to the pile or by turning it. Heat is a natural byproduct of the microorganisms' own action, and is raised in a two-phase process. First, the mesophilic (moderate-temperature) phase, which can last a couple of days, brings the temperature up to a threshold (approximately 44°C), whereby the thermophilic (high-temperature) stage can begin and the 'hot' or 'cooking' composting starts (approximately 45–65°C). Through active composting, this critical stage must be prolonged by turning or aerating the pile for up to two weeks. This will kill most pathogens and seeds without taking out the vital nutrients needed for growth. With no further intervention the compost will cool down, this stage will end, and the microorganisms will revert back to the mesophilic stage to further mature the compost and decompose the matter. Your compost can be considered finished, or 'cured', when the organic matter has turned to a consistently dark brown colour. This can take as little as six to eight weeks or as long as two years, depending on the maintenance (active or passive), the ingredients and the system.

There are many ways to design a compost system to house all your ingredients. The smallest I believe you can make, if you have outdoor space (I will cover indoor compost options later in the chapter), is $1m^2$, which can be scaled up depending on how much space is available – you can never have too much compost! See The 'Diamond' Compost Bay System for how to build and use this design.

No matter what system you use, I recommend acquiring a compost temperature gauge to ensure your compost reaches the desired temperature (45–65°C). Through the art of balancing the carbon and nitrogen (brown and green) ingredients and regular turning of the pile, this should be easily reached. However, if you let the pile get too hot – over 70°C – you will kill the beneficial microorganisms within the soil, so be aware of this, especially in the summer months. You are also at the mercy of the elements, so if it's raining hard and frequently, then consider covering the pile; and if the pile is soaking wet, add more carbon or brown material. Alternatively, if you're going through drought, add nitrogen or greens to wet the pile, or soak it with recycled grey water or harvested rainwater.

I've created many different compost piles and systems through the years, some good and some bad. But I've found that if they are managed well and correctly, they soon become the lifeblood of the garden and its most important function. Once a good aerobic system is in place, the labour involved will be far less daunting, especially when the knowledge of its benefit is understood. The traditional misconception of composting – a pile left in the corner with old carpet thrown on top – is slowly fading away, and as we become increasingly aware of the microscopic functions at play, the more we will reach for the pitchfork with renewed motivation and purpose.

THE 'DIAMOND' COMPOST BAY SYSTEM

I devised this system as a 'small-scale' alternative to the large open bay method. With the Diamond system, 'hot' composting is actively encouraged, its simplified design aiding both the turning and management of the pile needed to heat it to the required temperature. It also uses only durable reclaimed and recycled materials, which adds rustic charm to the garden and will last for many years to come. With a surface area of only 1m² it is adaptable to the smallest of gardens and is deconstructed and rebuilt with ease, should you need to move it.

1. Source some recycled scaffolding planks from a reclamation yard or directly from a scaffolding company.
2. Mark out and cut three 1m lengths and six 0.475m lengths from the planks. Cut at a 90-degree angle to get the straightest edge (figure 1).
3. Using angle brackets, fix two of the shorter planks to the centre of each longer plank at a 90-degree angle. For added strength, use at least 100mm-length brackets (figure 2).
4. Find a suitable site and place the first section on level ground. Stack the next section directly on top and repeat with the last. Add stakes, dug into the ground and screw-fixed to the boards if needed (figure 3).
5. You can add a panel roof in the winter if your pile is getting too wet.
6. Now you're ready to start composting (figure 4).

4

Before getting the pile going, I advocate saving back a decent amount of nitrogen and carbon elements. Chop and cut down as much of the organic matter as you can – especially woody ingredients like roots, branches and twigs – then begin by roughly layering a series of consecutive brown and green layers on top of each other. What you're trying to achieve is a good aerated pile, mainly through brown material, but with plenty of wet green matter to create enough heat. The stages for turning your compost in the Diamond system are as follows:

Stage 1. Fill up the first bay with fresh compost until full, then turn this pile into the adjoining bay (clockwise or anticlockwise). Use a pitchfork to turn the piles, breaking it up as you go (figure 5).

Stage 2. Again, fill up the first bay with fresh compost until full, then turn the pile in the second bay into its adjoining (third) bay and the fresh compost into the second.

Stage 3. Once again, fill up the first bay with fresh compost until full, then turn the pile in the third bay into the final bay. Repeat the turning process as before with the other piles.

Stage 4. Finally, fill up the first bay with fresh compost, and every bay should now be full. Begin taking the now-matured compost from the final bay and repeat the process from stage 3. Figure 6 shows different stages of compost, from fresh in the bottom left quadrant, anticlockwise to finished in the top left. Use a temperature gauge to track the temperatures of the piles as they traverse through the bays.

5

6

Leafmould and Loam

Compost will always benefit from utilising all the very best ingredients, but it can also be advantageous to make separate 'one-source' compost piles to increase the availability of unique soil structures, and leafmould is one of those. Leafmould is a dark, crumbly compost made solely from dead leaves. While it doesn't have the nutritional value of standard compost, it can be perfect for raising seeds or cuttings, which don't need an overabundance of nutrients. You can make leafmould using the same type of bay or bin as for traditional compost; just remember to turn the pile and keep it damp. Your compost will be ready to use by the next time the leaves start to fall.

Leaves turning to leafmould in a recycled bathtub.

Another one-source compost to consider making and using is turf. Loam-stacking or turf-stacking is the art of turning any dug-up lawn or turf face down, then laying more on top in a mound or stack and leaving it all to decompose for a good one to two years. This is the traditional method to create loam, or fibrous soil, and produces a superb structured soil that already possesses your land's unique microorganisms in it – a great advantage. Loam made this way is also an excellent addition to a potting mix for more binding and long-lasting soil.

Bokashi, Compost Teas and Biochar

While compost is primarily for the benefit of the garden, it would be remiss not to see how it helps us reduce the amount of waste from our homes. As mentioned earlier, in theory, anything of organic origin can be composted, though some materials will take much longer to decompose or require more effort and maintenance – cutting down branches, etc. It may seem obvious why sending less waste to landfill is hugely beneficial, but it's important to understand how conventional waste management correlates to composting. Landfill is a crude form of anaerobic composting, with no regard for the ingredients and therefore completely ineffective, producing tons of methane gas that contributes to the climate emergency. When we compost at home we too are releasing gases – aerobic composting produces carbon dioxide and ammonia – although this is only natural as it's a replication of the ongoing process in the natural world. Anaerobic composting occurs in nature too. However, unlike landfills that release gases into the atmosphere, in the natural world these gases are contained by either soil (sediment strata) or 'locked in' to the water over many years in bog or marshes. This is important to understand, as it shows that when done incorrectly composting can easily contribute negatively to the environment.

With this in mind, why would we want to compost anaerobically, anyway? The main reason is that cooked kitchen waste, like meat, fish, bones, or dairy can attract rodents and vermin to aerobic 'open' compost piles. So by sealing these ingredients either underground, underwater or in a container, we can avoid that outcome. It's best to think of anaerobic composting at home as a contributor to one's main composting process, not as a completely unrelated method; once you've anaerobically composted your ingredients, they can be added onto the main compost pile. With this in mind, let's consider what types of anaerobic composting can be done at home and how to use the results.

MAKING AND USING A BOKASHI BIN

This simple and effective system uses affordable and recycled materials in a design that is easy to build and use. Upscale to larger buckets, dependent on the amount of waste you produce.

1. Source two used 15–20l plastic buckets with sealable lids. You can find them through farm, food or building trades.
2. Using a 6–8mm drill bit, drill approximately 15–20 holes in the bottom of one of the buckets. Be careful not to crack the bottom by drilling the holes too closely together (figure 1).
3. Place the bucket with the holes inside the bucket without, and begin filling with food waste. Add the organic waste in batches, rather than singular items, in order to create layers (figure 2).
4. Then place a layer of inoculated bran over the top and seal the lid shut until you have more waste to add (figure 3). Inoculated bran refers to flaky

organic materials – typically wheat bran – mixed with what are known as Effective Microorganisms (EM-1) and some sort of fuel for these microbes, such as molasses. You can make your own inoculated bran by raising your own EM-1 culture, or buy a branded bokashi mixture that already contains it. The former takes time to produce while the latter costs money – the eternal conundrum.

5. Once the bucket reaches a quarter full, take the top bucket out and collect the liquid or 'tea' from the bottom bucket. A tap can be affixed to the bottom to make this easier (figure 4).
6. Once the top bucket is full, set aside for two to three weeks, draining the liquid regularly to allow the waste to ferment. Then take the waste and add to an existing compost pile. The liquid is an extremely potent, but acidic, inoculated natural fertiliser that can be diluted to a ratio of 50:1 (water:concentrate) and applied directly to the plants or soil.

3

4

Bokashi

Bokashi is a type of dry anaerobic fermentation that includes among its ingredients the added input of specifically selected microorganisms, which convert the organic matter into a preserved or pickled substance. This process neutralises the organic matter and, in a way, pre-digests the food, making it ready for the standard aerobic compost method to take over. It couldn't be simpler to do, and the first step is to make the container that the food waste will be stored in (see Making and Using a Bokashi Bin on page 54).

Trench composting is an another more simplistic technique of 'dry' anaerobic composting that involves digging a pit or trench to fill with organic, uncomposted matter, which then rots beneath the soil. This technique results in the majority of nutrients being localised within the trench and therefore plants – traditionally crops like runner beans – must be sown and grown on top of the trench. There is nothing especially wrong with this method; it just has some rather obvious flaws. Firstly, you can't use any of the waste, like meat and dairy, as with bokashi, since it will attract rodents and won't break down quickly enough. Secondly, trench composting is only useful for growing plants on top of the trenches, which generally works with veg gardens and nowhere else, although it's perfectly feasible to bury waste and then plant a tree or shrub on top. In the end, it's far easier to add this waste to the compost pile, so you can use the matured compost wherever and whenever you need it.

Compost Tea

Anaerobic 'wet' composting, such as manure slurry pits or rice paddies, produce methane through their anaerobic bacterial activity. If this methane is not captured, it will release into the atmosphere, causing significantly more damage per unit than carbon dioxide. This is not something we want to be doing, but through adding just one element into the mix – oxygen – we will turn the 'wet' compost aerobic, and this then becomes what we know as compost tea.

The introduction of oxygen serves the dual purpose of alleviating all the problems and difficulties of 'wet' anaerobic

BREWING COMPOST TEA

This method is a simple and affordable way of creating the environment for brewing tea inoculated with aerobic microorganisms. When distilled to its basic elements, this seemingly complex process becomes easy and efficient to use again and again.

1. Drill holes 25–35mm apart in a flexible plastic pipe 0.75–1m long, then fit to an electric aquatic pump (one that runs on a stand-alone solar-powered battery).
2. Place this pipe in a 20l brewing barrel full of rainwater.
3. Take two handfuls of compost and a handful of garden soil, and add approximately 50ml of molasses. Then either add this straight to the water or place it in a filtering bag - muslin or stocking - and suspend it in the water.
4. Making sure the pipe sits at the bottom of the barrel, turn on the pump and leave it running for 24 hours.
5. After 24 hours, turn the pump off and remove the pipe. If the ingredients were put straight in, then strain the liquid into another vessel. Use the tea on the plants and soil relatively quickly (within hours) after brewing.

composting – higher risk of pathogens, release of harmful gases, time to ferment – while expanding the beneficial microorganisms population into the aerobic realm, where far more microbial diversity accumulates. Compost tea (and bokashi tea) also has the added benefit of being a liquid nutrient and microorganism feed, as opposed to a solid, like traditional compost. This means they can be applied directly to the soil or to plants with greater ease, regularity and concentration.

Biochar

Fermenting organic matter in oxygen-starved containers or under bodies of water aren't the only ways to derive additional nutrient benefits from materials that would otherwise go to waste; we can also use fire. But while burning brush, logs, branches or anything combustible is a fast and effective way to get rid of matter and material in our gardens, it also pollutes the atmosphere and is at odds with regenerative, ecological gardening. Yes, wildfires occur naturally – in fact some plants, like eucalyptus, require fire to propagate – but in the garden simple fires are wasteful and detrimental. Almost all organic material that was thoughtlessly burnt in the past can and should be composted, but if your land is accumulating an excess of wood, there are alternative methods and practices to mere burning that have become obvious and urgent in light of the climate crisis. Producing biochar is one such method that offers promise, with the tantalising prospect of added benefits to the soil that can last far beyond the span of our own lifetimes.

Biochar is the wonderfully evocative name for charcoal made exclusively for use as a soil amendment. Even though humans have made charcoal, and in fact used biochar, since time

Biochar's porous structure aids water retention and gives habitat to microorganisms.

immemorial, the idea can seem elusive to many, so let me simplify a little. When wood is burnt it transitions from its original state to charcoal, then to ash. It is an easy process to observe, one that we've all witnessed – a fire is lit, burns down to embers and these burn to ash. Making biochar is all about stopping this process somewhere in between the embers stage and the formation of ash. Why? Because these embers are pure carbon and the benefits of 1) not burning it and 2) putting it into the soil are manifold. Halting the burn prevents the wood's stored carbon from converting into airborne carbon dioxide, but it's what happens when we work this biochar into the soil that is truly special. Using biochar in this way sequesters its carbon into the soil, similar to the way carbon is stored as part of the natural, organic carbon cycle. However, the carbon in biochar will linger in the soil for far longer than most forms: hundreds if not thousands of years.

The process of creating and using biochar has been practised for centuries and even millennia, notably by the indigenous tribes of the Amazon, famous for their terra preta, or black earth. In the present day, farmers in the Upper Guinean forests continue to use similar methods, prizing the dark earths created by their ancestors above all others, and realising the benefits of their continued practice. These ancient tribes on both sides of the Atlantic added biochar to enhance the natural soils found in their respective regions, and the results have survived to this day, thousands of years after their original production. The prime reason for this incredible longevity (and the high fertility of these soils) is biochar.

More of biochar's complex properties and beneficial relationships to the soil are still being discovered, but in simple terms biochar's porous and absorbent composition helps soils retain water and nutrients, making them available for plants to take up. This porous structure also aids microorganisms by providing them with something to which they can bind, and better perform their function in accruing soil fertility.

Making biochar can be a straightforward process or it can be complex, requiring a higher degree of skill and involving more labour. The more complex method involves creating a kiln to

1
2
3
4

THE 'OPEN-PIT' BIOCHAR METHOD

The open-pit method for producing biochar is less labour-intensive and expensive than constructing a kiln, as it simply involves digging a cone-shaped pit in the ground and setting the fire within the hole.

1. Collect a heaped wheelbarrow load of dried and aged wood, then process it into relatively similar-sized pieces - each no bigger than arm's length (figure 1).
2. Dig a cone-shaped pit in the ground to a depth of approximately 50cm, then light a fire at the bottom of the pit. Use the dug soil to form a berm (raised mound) around the edge, increasing the pit's depth (figure 2).
3. Begin layering the wood on top of the fire in a criss-cross fashion. As each layer burns to ash, place another layer on top, completely covering the layer below. This criss-cross stacking method slows the burn down and reduces the level of oxygen (figure 3).
4. When the layers reach the tip of the cone, let the fire burn until the top layer shows signs of burning to ash (the amount of time this takes can vary, but is a matter of minutes, not hours), then douse the fire with water until it is completely extinguished. Figure 4 shows the point at which to douse the char with water; notice the white ash forming on the char.
5. Wait for the fire to cool down and then harvest the char from the pit, putting aside any pieces that haven't charred completely for another burn. Break the char down when harvesting to ensure all the fuel has charred (figure 5).

5

house the fire and slowly vaporise the wood by restricting its access to oxygen – a process known as pyrolysis. This can also be achieved by the more basic method of covering a woodpile with soil to seal the wood from oxygen. The fire is lit from within the soil mound and then left for two to three days before the soil is removed and the char can be extracted. Kilns produce the least amount of ash, as they are the most efficient at reducing oxygen and therefore combustion of the embers to ash. A small brick kiln can be made by building a 50cm wide by 50cm tall circular structure out of reclaimed brick and clay mortar (mortars are the binding between bricks and stones), then covering it with a cast-iron lid. It will need a small air hole at the bottom but otherwise should be mostly airtight to perform well. It takes a lot of skill to accurately estimate how long the wood needs to burn before it's ready to remove. A little easier to implement would be the open-pit method (see The 'Open-Pit' Biochar Method on page 61).

To transform from its inert state and become biochar, this char – whether produced by a kiln or the open-pit method – needs to be inoculated with microorganisms and minerals. Then it can be added back into the soil. Otherwise, what you have is, in effect, charcoal, which when added straight will imbalance the soil's complex tapestry. Over time, adding simple charcoal won't be a problem, as the char becomes naturally inoculated in the soil, but this may take years. So to speed things up, we can add the char to the compost heap and let the inoculation take effect as the compost pile matures. When the biochar is then finally added to the soil, through the compost, its full benefits for the longevity and health of the soil can be realised.

Whenever I make biochar, I always have the sense that I'm privy to some ancient knowledge, harnessing fire for a useful function like never before. I've long admired the practicality and versatility of fires, especially campfires for cooking and warmth, where the type of wood and stacking method can change to suit the situation. But fires in the garden have always felt rushed and uncontrolled, a job or task to be done as quickly as possible. Over the course of a year, we tend to accumulate scattered wood into a pile and then decide to burn it all when the pile gets too big. In large private estates and public gardens, this can be even more frequent and the fires monstrous affairs as big as bonfires. However, burning for biochar is more interactive; the fire is never allowed to burn beyond your control and its energy utilised and reserved. The open-pit method doesn't take long, so at no point can you begin to gaze into its hypnotic flames or bask in its warmth. Fire is a primitive technology but that doesn't mean its purpose or value is less than any other; on the contrary, we need to learn more about its possibilities and realise its wealth of untapped potential.

Through the Body

As we've covered, human-managed composting replicates processes that exist in the natural world; but there is one other composter found in nature that is just as effective to harness for the purposes of the garden: the stomach. Whether human, animal, bird or invertebrate, digestive systems produce organic composts that are full of the nutrients and elements essential to soil health. Organisms, people very much included, turn food to compost almost in perpetuity, and learning how to use this precious manure resource is fundamental towards making a garden in full organic health. Farmers all over the world have been using the manure from their animals to fertilise crops since time immemorial, but in gardens of all shapes and sizes – whether as small as a pot on a window sill or as large as an orchard of a couple of acres – this same principle can be adapted and applied. What matters is that our plants benefit from compost and healthy soil, and digestive systems – whether animal, or even ours – can be the perfect way to achieve this.

Vermiculture

One of the most effective proponents of the digestive composting method are the princes of the soil: worms. If you live in a flat or only have a small balcony, then vermiculture – the cultivation of worms – may be your best composting option. Even in the smallest garden I would advocate an aerobic compost system above all others, but many people with a love for plants don't have that luxury and in such cases vermiculture is the saving grace. Once again, it's essential to understand what vermiculture is and what it's hoping to emulate. Vermiculture is a form of cold, aerobic composting. While sustaining a constant high temperature is key to a traditional aerobic compost pile, vermiculture employs the opposite strategy; we want it to stay at a steady low temperature for as long as possible. When a standard 'hot' compost pile cools down, and before it heats up, worms find their way into it and do exactly what vermiculture seeks to replicate. All vermiculture does is make the process intentional, whereby the gardener adds hundreds of these worms at the beginning of the composting. This enables the worms to digest the majority of the waste instead of the microorganisms, with the byproduct – their worm casts, or excreta – being the resulting compost. This entire process can happen in a dark box in the kitchen, hence why it's the flat-dwelling houseplant enthusiast's best friend.

Worms consume organic matter and excrete casts – vermicompost.

In a complex world, simplicity can be a guiding light, and it's the same with vermiculture. You really only have two choices at the home scale. The first is a box with a lid and seeing as this doesn't have to be airtight, you can make one from recycled wood. The second choice is a stack or tower of boxes with a lid, also best made from recycled wood. The reason for the stack is for ease of harvesting the casts. The one-box method requires removing the worms when you take their casts out. The towers,

Table 2.2. What to Feed (and What Not to Feed) Your Worms

Good for Worms	Bad for Worms
Cardboard, paper and straw (added as bedding)	Animal fats
Coffee grounds and tea leaves	Citrus fruit
Cooked food (plant-based)	Cooking oils
Crushed egg shells	Dairy products
Fruit and veg scraps	Meat (cooked or uncooked)

however, work by allowing the worms to gravitate to a higher stack when they're finished with one; therefore, when you take out the lower stack, the casts are already free of worms.

As for what type of organic matter to fill the box or stack with, this doesn't have to be complicated either. Just start with a good thick layer of dry carbon – ripped up newspaper and cardboard will do – then add some soil from a mature compost pile, and then the worms. You can harvest the worms from an existing compost pile or order them in, but in either case they will invariably be red wigglers, or brandlings. The worms will then need food, and this is where things get slightly more complicated. Vermicomposting won't process all of your organic waste, only a section of it, and what goes into the vermicompost needs to be cut and chopped up to be most effective – see table 2.2 for what can and can't go in. It is your role to shepherd the process and the same mindset of conventional composting applies: if the matter looks too wet, add more carbon; if the matter looks too dry, add more nitrogen. Diligence is the key and the process can only go wrong if the chef isn't tending to the pot.

Excreta Eccellente

We humans, just like most other living things, excrete highly nutritious waste that, used correctly, can be composted to turn into sweet-smelling, rich organic soil. The only thing we need to do is to unplug from the sewage pipe and create a composting toilet or collection.

BUILDING A VERMICOMPOST STACK-HOUSE

1. Source some wooden pallets and begin extracting the boards, making sure you don't damage them in the process. Loosen the boards with a hammer and wood chisel, then pry them off with a crowbar (figure 1).
2. Process nine uncut full lengths (70cm) and six shorter lengths (30cm), sand down and put to one side. Then cut twelve 45mm × 45mm softwood stakes at 100mm lengths.
3. Using 40mm screws, fix two of the long boards (70cm) to two of the short boards (30cm) in a rectangle. Then fit the stakes at each corner, halfway up the board, leaving 50mm proud at the bottom. Repeat this process twice more. Pre-drill the screw holes to avoid splitting the boards (figure 2).
4. Using U nails, secure small-mesh chicken wire to the bottom of each tray, then fit together. Place one tray into the ground and then slot the next tray on top of this. Repeat for the final tray. Each level should fit flush with the next, sealing the whole unit (figure 3).
5. Construct the roof with the remaining three full-length boards, held together by screwing two 20mm × 35mm batons of no more than 30cm length across their width. Using the same-sized baton, construct a 70cm × 33cm rectangle (screwed or nailed at the

1

2

corners) and attach this to the boards with two hinges. Place the roof on the top. In figure 4, the top level is ready for the worms to gravitate from the middle layer. The roof keeps the worms in and the stack-house dark (figure 5). Add an impermeable collection tray under the stack-house if you wish to harvest the nutrient-rich worm leachate.

3

4

5

This method, much like many in the art of composting, is not anything new and has its origins in the days before the invention of the sewage system and the mass production of the flushing toilet. The fertilising properties of human excreta have always been known, and 'night soil' was once commonly taken straight to the farmer's field. Hindsight tells us that the problem here was the fact that our forebears didn't quite understand how bacterial pathogens can prevail in the excreta and then be absorbed and contained within the crop. This led to the sewage system and the removal of pathogens by chemical means – a logical solution apart from the fact that such a system requires enormous amounts of energy and freshwater, and bypasses the knowledge that pathogens can be killed by the correct practice of composting. 'Humanure' can be made safe by applying the same simple science behind aerobic, 'hot' composting, whereby temperatures of 55–70°C are reached and maintained for a certain period of time, killing all pathogens in the process. (Introducing water – as in slurry pits, for example – would initiate anaerobic, 'wet' composting by which pathogens would only truly be excluded after a long period of time – at least two years – if at all.) Despite this fact, fear of pathogens remain in the modern population, where the composting toilet is seen as an unsanitary relic of the disease-ridden past.

To create a compost toilet, you will have to undertake some quite radical re-plumbing, but it's certainly not beyond the realms of reality. The main question concerns the quantity and whether your composting pile can sustain the added amount of organic matter, but let's assume for now that you can. The next thing to consider is whether you have enough room for a functional composting toilet within the infrastructure of your bathroom or house. Or whether you are just collecting the excreta to take to the compost pile in the garden – much like the system for food waste. The former is quite engineered, so let's consider the collection method. All the same composting rules apply: you'll need a collection vessel or bucket, air via a vent for aerobic composting, carbon material (sawdust, shredded paper or cardboard, etc.) and nitrogen – our waste. That's it. The infrastructure isn't important; it's the management. If the waste smells bad, that's because you've created anaerobic conditions and your systems needs more air, which means you need to either bring more in via a larger vent or add more carbon. If the waste has excess moisture, then, the same again, you need to add more carbon. If you layer carbon on the bottom of the vessel and then after each addition of excreta, this will create a good aerobic environment. Remember, even if you've got a composting toilet, the excreta still needs to go through a 'hot' compost process after collection to kill pathogens. Without doing this, the whole endeavour fails.

Animal Manures

If we are efficient compost producers, then our animal cousins and feathered friends are the true masters. Horse, cattle and chicken manures are some of the very best compost ingredients. Farmers already know this only too well and it can be harder to get hold of cattle or farmyard manure than horse or chicken manure, as they use most of it to fertilise their own crops. There is a slight dilemma for the ecological gardener, however: the majority of cattle manure is not organic and can contain antibiotics, pesticides, hormone treatment and other synthetic elements. Yet not to use this excess waste doesn't seem to make sense either, especially if the process of 'hot' composting and integration back into the soil transforms these elements back into organic, healthy compounds. The same issue arises with human excreta – was everything you and your family consumed organic? Did you take paracetamol? Did you smoke or drink alcohol? My opinion is that we will not turn the world organic overnight, and this is all part of the regenerative nature of looking to rebuild anew, leading a light to more natural, holistic paths. Composting is integral to that and shouldn't be divided into good and bad; it can only be done correctly or incorrectly.

With that in mind, how do we get hold of animal and poultry manure? And what shall we do once we have it? If you live in a rural area, then getting a delivery of cattle manure from a friendly farmer is an option, but it's far more likely that your source will either be a stud farm or stables or by keeping chickens yourself. With the horse manure, I recommend visiting a local stables and just asking if you can have some. Usually they are more than happy to accommodate your wishes, as they have more than they know what to do with. Keeping your own chickens will always be of enormous benefit to your garden, and collecting their manure is a simple process of using carbon bedding – straw or wood chips in the coop – and clearing it out and onto the compost heap. Cattle and horse manure (which will already contain straw bedding) need to go through the 'hot' composting process to eradicate pathogens and seeds. If you are taking delivery of a fresh load of either cattle or horse manure, then I suggest setting this aside in a separate pile for at least a year, then slowly adding it into the main compost in stages. Fresh manure needs time to mature in order to reduce its acidity and allow microorganisms to break down harmful pathogens and unwanted chemicals. To be extra safe, seek out only organically produced manure.

Once you have mature, well-rotted compost, then putting it out in the garden and using it is the easy part. It can go on the vegetable beds and flower borders as a spring mulch, and at the base of newly planted trees and soft-fruit canes and bushes. It can also be used as a potting mix for raising seeds and seedlings, as well as divided perennials or cuttings. And it can be used in containers and pots for displays in the house and around the garden. In almost every facet of the garden and the gardening craft, compost is a virtue and valued asset. How much you need and how much you make is all dependent on your individual situation, but if you are a keen grower of living things, then you can never be without enough compost.

The benefits are myriad and spectacular. Adding compost will add all the key nutrients and elements needed for growth and renewal to generate and sustain life, including nitrogen, phosphorous, potash, magnesium, calcium, iron, carbon, zinc, boron and more. It will balance pH levels to stop a slide towards saline or alkaline soils, retain water in sandy soils and break up heavy clay. This contributes to an increase in natural water filtration and raising of fresh-water levels in our aquifers and river systems. It will also hold and support invertebrate and insect life, which in turn will provide a food source for the birds and mammals that rely on it for their survival. But above all, it will introduce more microorganisms into the soil than anything human invention can dream of recreating. Almost all composts that can be bought at a store or by delivery will have been heat-sterilised to a lifeless medium – killing all the beneficial microorganisms in a pathogen-induced panic, like a drugged-up patient on antibiotics for a common cold.

Young fruit trees need plenty of nutrient-rich compost until established.

In taking the organic waste that we and the natural world produce and bringing it back to the soil through composting, we mustn't underestimate the importance of the role we are undertaking. In stepping into the ecological cycle, we will contribute carbon dioxide into the atmosphere as a result. However, the more plants we grow and produce in a healthy natural way, the more carbon dioxide will be drawn down from the atmosphere. Never has it been so critical to understand this and all other natural processes at play in the garden. If we are committed to good organic gardening methods and diligent stewardship, we can enable our gardens, and the soils within them, to be vibrant ecosystems, offering sanctuary and abundant health to all organisms that dwell within them and upon which all life depends.

CHAPTER 3

Plants

Plants are the principal attraction in any garden and the sole reason many of us are drawn to the art and craft of gardening. Their infinite variety is a constant source of fascination and curiosity, and to learn more about their mysterious ways is to tap into the gathered wisdom of the natural world. As ecological gardeners, we have the opportunity to return land to its wild origins, and the plants we use and choose are vital towards this regeneration. Every garden is part of a wider environmental biome – here in the UK, temperate broadleaf forest – with each country within these biomes having its own indigenous fauna and flora, distinct and unique from those of others. Therefore, to think about the wild origins of plant life in the UK is to think about its indigenous species. Utilising indigenous species means that your garden will be better adapted to its environment than one full of non-native species – one that possesses more defenses against the ravages of a changing climate, and also provides more for local wildlife. Indigenous plants offer no less in terms of beauty, function or provision than their exotic counterparts either. On the contrary, we need only to realign ourselves once again with the plant heritage of our

Plants are the lifeblood of our planet, drawing down CO_2 and giving out oxygen.

surroundings to behold the treasures held within.

Ecological gardening isn't just about composting or using recycled materials; it can also be about relaxing some of the more compulsive gardening tendencies of tidiness and cleanliness. Wandering wild-flowers know no boundaries, and the more you satisfy their needs the more the diversity of your garden will explode. You can also create dedicated areas that gently merge with the wild, like wildflower meadows and hedgerows. And if ever there were one category of plant to use to readdress the decline of our ecology and mitigate the broiling climate, it is trees. Their great green lungs are the engines of renewal, regeneration, and the rebalancing of our lives on Earth.

Indigenous Plants

Plants play a role in both the natural and cultural history of every country in the world. From Cuba (butterfly jasmine) to Australia (yellow wattle), many nations designate national flowers to proudly extol their indigenous flora. Here in the UK, the national flower of England is the red rose, ironically a non-native species, contributing not only to an illusion of what is native to the UK but also a chequered and lettered history of horticulture. Admittedly, defining what is and isn't native to the UK can be tricky; the amount of geological time we've had for our plant communities to establish themselves is much shorter than many other countries – only 12,000 or so years since the retreat of the glaciers at the end of the last ice age. And since that relatively clean-slate beginning, our history has been one of floral flux, encompassing the first migratory farmers and growers; early Roman kitchen and pleasure gardens; the Victorian plant-hunter boom; and the modern-day horticultural world. All this has changed the tapestry of wild plants in this country to such a degree as to render the native and the new almost indistinguishable. Some might say this is the way of the world and we should let it be, but some introduced species can affect and alter the indigenous flora and fauna in a variety of negative ways and this practice shows no sign of stopping soon.

Today we plant more non-native species in our gardens than ever before, while farmers have more advanced chemical methods to eradicate the native flora in the pursuit of crops or livestock. The habitat for our native species is similarly decreasing at a rapid rate as we ourselves increase in numbers, squeezing the band in which they take refuge into ever smaller fragments. It would seem to be a perfect storm for the demise of our native species, unless we take it upon ourselves to turn the tide and offer sanctuary for them in our gardens. Not only are these the plants of our country and place, but also they are far more robust and long-lasting than the exotic cultivars that dominate the gardening market – annual bedding plants, frost-prone perennials, etc. – and crucially the easiest to maintain with the smallest of energy inputs. But above all, they are the lifeline of native and migratory wildlife. The symbiotic relationships between flora and fauna have been forged over millennia; when plant species die out, wildlife diversity declines accordingly, and the overall health of the ecosystem suffers. By planting indigenous plants in our gardens, we can help reverse their decline in the wild and restore the elemental natural balance.

Choosing to grow only indigenous plants doesn't make the planting of a garden any less subtle or refined, as our goal remains a garden full of plants that will thrive and be admired for their beauty. To achieve this, plants need to be placed not where we desire but in situations that most closely resemble their habitat in the wild. All gardens contain a series of microclimates within them due to different levels of sunlight, heat, shade, rainfall, wind and frost. So if we can find plants that we know thrive in these climates in the wild, planting them in a similar context in the garden will give them the best chance of success. This is the de facto template by which all the plantings of the garden should adhere, and in doing so the style of the garden will present itself naturally. In every situation or microclimate that a garden possesses, we will find indigenous plants that will prosper and inspire no less admiration than a cultivated equivalent.

Open

In the areas of the garden that receive the most light, some indigenous species will thrive and produce an abundance of form and colour throughout the summer months. All these plants will tolerate a period of drought and prefer a well-drained soil. Recommended species include:

Musk mallow (*Malva moschata*). By far the most beautiful mallow, this species loves the open.

Ox-eye daisy (*Leucanthemum vulgare*). One of our most favoured national flowers but no less of a beauty for it.

Viper's bugloss (*Echium vulgare*). Will tolerate parched conditions, and its blue flowers are a welcome riposte to an overabundance of pinks and whites.

Field scabious (*Knautia arvensis*). Features pincushion-shaped, lilac-coloured flowers that fall and nod about with a singular elegance.

Ox-eye daisy is also known as the 'moon daisy' for the moon-like glow of its flowers.

Shady

Light shade must be the most ambiguous of all the conditions, so it can be easier to think of a woodland edge – where the light allows plants to thrive that would otherwise be smothered by the darkness of a thick wood canopy. In the garden, it's reasonable to assume you will have some elements with height and therefore will cast shade. Light-shade plants don't need constant dappled shade; in fact, they tolerate a wide spectrum of conditions from a lot of light to a little, just not the extremes of full shade or constant, drying sun.

Cow parsley (*Anthriscus sylvestris*). An all-time hedgerow favourite.

Red campion (*Silene dioica*). More magenta than red, this species will flower for months on end.

Great burnet (*Sanguisorba officinalis*). Possesses a wonderfully architectural elegance and is, like many others, preyed on by an endless succession of cultivar cousins.

Common valerian (*Valeriana officinalis*). Criminally overlooked in favour of other species.

Old names for cow parsley include Mother die, kex, kecksie, Grandpa's pepper and Spanish lace.

Dark

Darkness is the ultimate replication of the broadleaf woodland setting and a window back to possibly the most ancient of all our indigenous species. These plants that tolerate high levels of shade won't thrive in perpetual darkness – although the ferns will not mind this – but adapt to make the most of the changing seasons and the subsequent levels of light that this brings.

Green hellebore (*Helleborus viridis*). A more refined and elegant choice than the dizzying number of hybrids available.
Primrose (*Primula vulgaris*). Any number of the horrible cultivars can't hold a flame to it.
Wood anemone (*Anemone nemorosa*). An ancient woodland indicator species. Common names include the brilliant misnomer 'wooden enemies'.
Bluebell (*Hyacinthoides non-scripta*). Makes use of the available light in spring if placed under broadleaf trees. Not to be confused with the invasive Spanish bluebell (*Hyacinthoides hispanica*).
Hard fern (*Blechnum spicant*). Will tolerate the poorest levels of light throughout the year.

Surely the bluebell is a greater candidate for the national flower of England than the rose?

Biting stonecrop. Old name: welcome-home-husband-though-never-so-drunk.

Dry

Whether a garden is newly formed or mature, gardeners often have a strange propensity for believing that hard-standing features like walls, paths and terraces are entirely separate from the plants they are set among. Not so. If you construct such features with natural materials and methods, these areas can be pockets of opportunity for plants that flourish in similar circumstances in the wild, among rocks, sand and gravel.

Biting stonecrop (*Sedum acre*). Will happily take up residence in the crevices of walls.
Wild chives (*Allium schoenoprasum*). Likes to colonise the cracks between flagstones and wall stones.
White rock rose (*Helianthemum apenninum*). Flowers between May and July.
Procumbent pearlwort (*Sagina procumbens*). Takes hold in small crevices.

Grassland

Grasses are an integral part of any wild planting and, as I will expand on in more detail later in the chapter, I believe a hay or wildflower meadow to be an essential element in regenerative, ecological gardening. Although many of the grasses in meadows are small and repressed, numerous other species grow tall and free, adding structure and drama to your plantings and encouraging a wider scope of wildlife habitat.

Tufted hair grass. Conspicuous in the midsummer with its billowy habit.

Purple moor-grass (*Molinia caerulea* subsp. *arundinacea*). An absolute favourite of mine, glows like fiery embers through the autumn.
Tufted hair grass (*Deschampsia cespitosa*). Another fine native, which is better

suited when given space to expand and billow out, as it likes to do.

Wood melic (*Melica uniflora*). Offers a contrast to taller grasses with its elegant and dainty habit.

Bog

You may find there are areas in your garden that are constantly damp or liable to hold water in the winter – repress the urge to fight against this. These types of areas are rapidly diminishing in the wild, with untold effects on the environment and the natural world as a result. Britain has many plants adapted to this condition, being a historically boggy landscape, so by planting species that are threatened in the wild, we are not only regenerating a falling group but also being given the chance to revel in the beauty that we seldom see.

Ragged robin (*Lychnis flos-cuculi*). Hands-down one of the prettiest of all our native plants.

Meadowsweet (*Filipendula ulmaria*). Sweet-smelling and found mainly in damp places.

Hemp agrimony (*Eupatorium cannabinum*). A victim of the cultivar kaleidoscope, so seek it out over all other forms.

Betony (*Stachys officinalis*). Dynamite little late-flowerer.

Southern marsh orchid (*Dactylorhiza praetermissa*). Reflective of our national blindness to native orchids; they are tricky to grow from seed but will be fine once established.

Ragged robin. Ragged and crumpled, and all the better for it.

In beginning your journey of discovery with these indigenous plants, you will undoubtedly find some harder to acquire than others. However, as with most things in life, the extra effort will be outweighed by the fruits of your labour.

In the main, you will have three options laid out before you for procuring indigenous plants. First, acquiring seed, whether you purchase it or are given a collection from a friend. Either way, be judicious to make sure the seed is truly native and not of the later introduced varieties. You must then either raise this seed as seedlings – best if you want to produce groups of plants for specific areas – or just broadcast (scatter) the seed where you wish the plants to be and hope for the best. Second, you can buy what you can in plug form – juvenile plants in a small amount of soil in trays of plugs or mini pots. This strategy is useful to avoid the sometimes hit-and-miss nature of growing from seed, but the range of species you can purchase this way will be much more limited. Finally, if you want to bypass all plant husbandry, then you must look for grown species in larger pots that you can transplant straight into the ground. However, remember that as you diverge further and further from traditional plant buying (i.e. garden varieties and cultivars), native species will become increasingly harder to find.

It never ceases to amaze me how infrequently I've seen native perennials, biennials and grasses in the gardens I've worked in throughout my career. Yes, there are plenty of trees and a fair selection of shrubs, but few indigenous perennials, biennials and grasses. This is partly the fault of public gardens and parks, which give garden owners their inspiration, and carries down through to the nurseries and garden centres. If indigenous plants aren't available at any stage, how will anyone know what they are even missing in the first place? Add to this the rapid assimilation of native habitat under the plough and it's little wonder most people don't even know what a ragged robin looks like, let alone that it's part of our indigenous flora. I believe if we were to see our indigenous species more, like we do with oaks or yews, we would take them to heart as much as any other species and be compelled to put them in our gardens not by necessity but by choice.

Wildflower Meadows

There can be little doubt that the surge of interest in wildflower meadows in recent years is due to a growing appreciation for indigenous plant species. Many have lamented their almost total loss, but many more have had to ask: What is a wildflower meadow? The answer would seem at first rather simple, with many of us able to conjure up an image of a colourful rolling pasture. And yet, as with much in the mysterious world of horticulture, everything isn't always quite what it seems. Meadows can take on many novel forms and although they can sometimes look similar, they have vastly different means of survival.

The most distinctive difference is between those that are natural and those that are manipulated by human and animal intervention. Natural meadows are areas of ground that usually experience harsh climatic conditions that resist the normal successional formation of forests. They might occur above the treeline in the mountains, along the salt- and wind-lashed coastline, or in areas of severe drought and fire such as deserts or prairies. These specific environments – sometimes called perpetual meadows – are the perfect places for perennial grasses and wildflowers to make their own and, over many thousands of years, create highly diverse sustainable habitats. Natural meadows can also form as successional stages where a previously forested area suffered a disturbance – a fire, for instance – but these will only ever be fleeting and are sometimes called transitional meadows. All other meadows that we are likely to see were formed by human intervention.

It all begins with agriculture, and the way our early farming ancestors, by chance or design, managed to marry these two distinct types of meadows together. First, they created a transitional meadow by clearing the land of trees and scrub through burning or felling, and then mimicked the perpetual meadow by establishing grazing patterns for their livestock to stop the forest

A coastal meadow prevailing where other plants cannot.

ever forming again. Why hasn't the type of grazing still practised the world over today – in a far more intensive form – not resulted in a world full of wildflowers meadows, then? Because wildflower meadows require a delicate balance between over- and under-grazing. Traditionally, another stage of agricultural production, hay-making, helped strike this balance. To feed and sustain a population of livestock year-round on grass, the traditional method was to allow some fields to go ungrazed through the spring and early summer, and then cut in late summer for hay that could be fed through the winter. This would give perennial flowers and grasses the chance to reach their potential within the year, and as farmers left these fields like this for many hundreds of years, they became, in essence, managed perpetual wildflower meadows. The key to the meadows' species diversity lay in the fact that this method depletes the soil of nutrients due to the full growth of the plants, much like growing crops would; therefore, over time, grasses began to yield to more adventitious wildflowers that fared better in the nutrient-poor soil. Unfortunately, farming practices changed completely after the Second World War, with the introduction of herbicides, faster growing grass species and silage production, all combining to eradicate the late summer hay-cut, and, as such, meadows disappeared in line with these changes.

This history is vitally important before we embark on creating our own wildflower meadows, and so too is one further distinction. Meadows are always perennial, but they have a close relative in annual cornflowers. Once again, agriculture laid the path for their prosperity by producing arable land for crop growing. The soil is first ploughed to remove all vegetation, and this bare earth induces the flowering of a vast seed bank that has been waiting patiently for its moment to break out of dormancy. As this cyclical ploughing persisted, so too did an ever more diverse population of annual species benefiting from the lack of competition from their perennial cousins. The farmers saw these species as only weeds, taking away from their designated crop, and once again changed their methods, resulting in the disappearance of arable wildflowers. We need to realise this difference, as an annual cornflower field requires annual ploughing to induce the seed into flowering and is not a meadow at all. This is counter to the methodology for creating a wildflower meadow, which requires absolutely no ploughing whatsoever – an important distinction to remember!

If understanding the difference between wildflower groups is vital, this pales into insignificance beside knowing what a wildflower meadow does for the natural world and ultimately why we would want to restore and recreate this habitat. Meadows offer a wealth of sustenance and habitat for wildlife, but perhaps more importantly they are evocative and beautiful to us. Much of wildlife habitat can be rough and scraggly, but wildflower meadows are like paintings of loose expression and colour, which, combined with what they provide to the natural world, makes them feel like the perfect blend of human and wild creation.

Making Meadows

Deciding which species to sow in your meadow is one of the first steps, and although this can be daunting, there are general seed collections to guide you. However, your site will always have variations of soil, aspect, water drainage and microclimate that are unique from any other, so it's worth researching the correct mixture of species suitable to your conditions. Also, in the centuries that wildflower meadows were present in the landscape, a

Meadow buttercups. Still an indelible part of childhood... do you like butter?

1

2

3

4

TURNING A LAWN INTO A WILDFLOWER MEADOW

The first thing to do is ditch the mower and let the grass grow long. You'll then be able to see which indigenous species already exist within the lawn. This will throw up some surprises - good and bad - and it's critically important to make a note of what is there. It's more than likely that grasses will be extremely dominant, and if we left the meadow at this ratio, we would only ever be growing long grass. So when the summer ends and autumn falls in, we need to take the lawn to task and start turning it into a wildflower meadow. The reason for sowing in the autumn and not the spring is that most of the species release their seed at this point naturally, and some even need a period of winter cold to break their dormancy. It also increases their chances of developing a seedling before the dominant grasses take hold in the spring.

1. In the autumn, cut the lawn to the absolute lowest your mower or scythe will allow. Go over the lawn a few times, lowering the height after each pass (figure 1).
2. Depending on the size of the area, using either hand or machine, scarify or harrow the lawn until 80 per cent of the existing grass is ripped out.
3. Rake this up and remove. Expose bare soil in which seeds can germinate (figure 2).
4. Source a 100 per cent (no grass) indigenous wildflower seed mix of perennial species and sow at the recommended rate (figure 3). Mix the seed with sand so you can see where you've sown (figure 4).
5. Press the seed firmly into the ground by walking over it, using a weighted roller or tracking over it with vehicle tyres.

dazzling diversity of species took hold that became quite specialised to certain local regions. Many of these species are now either extinct or critically endangered, so seeking out the latter will not only restore a lost heritage but also engender a wider amount of biodiversity within the meadow. Common wildflower meadow species include: meadow buttercup (*Ranunculus acris*), ox-eye daisy (*Leucanthemum vulgare*), greater knapweed (*Centaurea scabiosa*), lady's bedstraw (*Galium verum*), meadow foxtail (*Alopecurus pratensis*), meadow brome (*Bromus commutatus*) and sweet vernal grass (*Anthoxanthum odoratum*).

Wanderlust will only take you so far though, and with meadows it's really only time and labour that will yield results. A wildflower meadow can be grown anywhere and be of any size, but why not start with the grassy sward many of us already possess: a lawn? A lawn is a type of meadow, albeit a rather strange one with unusual species in it, so it seems only logical to reimagine it as a wildflower meadow (see Turning a Lawn into a Wildflower Meadow on page 87).

Maturing Meadows

I wish I could say that creating a meadow is as simple as sowing the seed, but the truth is it isn't, and the likelihood of success still hangs in the balance. If your grass is still too dominant with species such as perennial rye grass (*Lolium perenne*), then this will shade out the growing seedlings and will need to be cut and all the cuttings removed. This is fairly common and can be combated in a number of ways. Firstly, plug plants can be grown from seed and planted in the meadow, giving the new species a much better chance of getting established within the dominant grasses. Or, more drastically, the whole lawn will need ripping out, impoverished soil put back in its place, seeds sown into this – again in the autumn – and the meadow started from scratch. However, before giving up the ghost too soon, we still have one ace in the pocket: yellow rattle (*Rhinanthus minor*). Yellow rattle is an indigenous species of annual wildflower found in most established meadows, and this is because it parasitises the roots of grass species. This weakens the grasses and helps wildflowers take hold. Although it doesn't need arable ground to germinate, it will still need to be given the space to go through a few cycles of seeding and growing to form an establishment. It will most likely be included in the original mix sown in the autumn, but it's so critical to the meadow that strips can be cut into the lawn and the yellow rattle sown into them to allow it space to develop and take hold.

A mature meadow where wildflowers are more dominant than grasses. Note the yellow rattle.

Assuming that your meadow has started life well, then its guidance will be of equal importance to further its progress towards a mature meadow. Remember, meadows require both nature and the human hand, and if we don't take part, the endeavour will fail. The first consideration is cutting the grass. In traditional hay meadows, this would've been done in late summer, but we want to create a habitat for wildlife and establish wildflowers over the entire flowering season, so ideally cut your grass every year in the early spring. However, since cutting closer to summer will reduce the vigour of grasses, an alternative is to start in late summer and slowly work your way out to the spring over the course of a few years. To cut, you will need a scythe or its more mechanical counterpart, the power scythe (only use if battery-powered). The scythe will cut the stalks and stems cleanly and neatly, making the collection of the cuttings far easier to move and handle.

These cuttings will have very important seed in them – especially yellow rattle – so allow them to disperse, or collect them before removing the straw. Many perennials will reproduce

Tips for Scything

1. Adjust the snath (scythe handle) to your body – move the lower handgrip to hip height, then place the upper handgrip the length of your forearm above this.
2. When mowing, spread your legs, bend your knees, keep your back straight and mow in a smooth arc with the blade always resting on the ground.
3. Hone the blade 'in the field' with a whetstone, then periodically sharpen the blade by peening it – a process of hammering the blade sharp.

The scythe is an ancient tool, but its practicality has not diminished over time.

vegetatively (cloning themselves off their roots), but most will also produce seed and move through the meadow this way, so if we collect the seed, it can be over-sown manually with the species we would like to encourage. Even with a good grouping of yellow rattle, it's advisable to sow more every autumn for the first few years as the meadow develops, to help tip the balance of grasses and wildflowers firmly in the flowers' favour.

After lots of continued hard work, an established meadow will take hold and the pleasures of visiting wildlife can be truly admired. Nectar-seeking insects such as bees, butterflies and moths will be in abundance due to the quantity and diversity of nectar-producing flowers, while mammals such as shrews, voles and hedgehogs will benefit from the added vegetation cover that enables them to hunt, nest and make tracks and trails. A whole host of other invertebrates including grasshoppers, leafhoppers and the larvae of moths and butterflies will also make their home among the stems. And birds too will come looking to feed on this abundant food source, with the possibility of enticing skylarks, curlews and meadow pipits to not only feed but also roost. In short, it will be a biodiversity hotspot with enormous benefit to wildlife and to particular species that used to thrive in this lost environment. It can only be left to ask, what is more beautiful: a lawn or a living meadow? There seems to be no contest.

Hedgerows

At first glance, a hedge would seem to be the very antithesis of an open, boundary-less, ecologically focused garden, as it attempts to enclose and restrict entry for the wildlife we seek to promote. Indeed, hedges are usually mono-species plantings only slightly removed from walls and fences. However, hedge*rows* are wildlife havens and biologically rich plantings that promote the use of native species as much as wildflower meadows. Think of them as mini woodlands, which in essence they are, their origins dating from the earliest clearances of forests for agricultural land, whereby a strip of woodland was left standing when all else was felled. These earliest hedgerows were used to demarcate the edge of property and hold livestock in place within the new pasture, as well as keep wild mammals out. It's possible that the thornier hedgerow species, such as hawthorn (*Crataegus monogyna*) and blackthorn (*Prunus spinosa*), were then added to the woodland strip for their obvious defensive capabilities and then some larger trees coppiced for fuel and firewood. This subsequent thickening of the strip formed a resemblance of the hedgerows we know today. This ancient

PLANTING A NATIVE COUNTRY HEDGEROW

So how do you go about creating a species-rich hedgerow in your garden? Firstly, you need to choose your species and then purchase the plants. They will be supplied as whips - one-year-old saplings - and will be delivered bare root in the winter and early spring, which is the best and only time you should plant a hedgerow (see table 3.1).

1. Acquire sapling whips, five per metre of the length of the desired hedgerow.
2. Dig two shallow trenches 30–50cm apart along the length of the hedgerow.
3. Plant the saplings at 40–50cm spacing within each trench, staggering the gaps between the two rows.
4. Spread compost mulch around each sapling and water thoroughly. Repeat the compost mulch in spring.

These saplings will grow fast and strong, and after a year or two I would then add some scramblers and ramblers as well as woodland bulbs and perennials (see table 3.1). The perennial species you choose will depend a lot on where you plant your hedgerow, as it could adjoin a wildflower area that receives lots of light or be next to buildings and predominantly more shaded. Regardless, the crucial thing to consider is making a hedgerow ecosystem rather than a hedge, as multiple layers of planting will attract the most variety of wildlife.

A hedgerow will support a vast array of wildlife.

A country hedgerow is open and sparse in the depths of winter.

heritage has been preserved in the landscape and its practice and craft carried on through the ages, albeit in slightly different forms and guises.

What at first functioned mainly for livestock became a lifeline for the woodland as the indigenous plant species clung on in a mazy patchwork across the barren grasslands. Wildlife too took refuge in this corridor and used its channels to traverse the changing landscape. As the hedgerows were ostensibly woodland edges, the extra levels of light increased development of biodiversity as species previously clothed in the dark canopy made the most of this newfound opportunity. Many of the ancient hedgerows also formed at natural delineations such as rivers, paths, ditches, dykes or roads, further increasing their unique microclimate and specialised ecosystem.

But what does this all mean for us in our gardens? Although hedgerows would seem to be countryside conurbations, we still have boundaries, and it's the hedgerows' ability, above all other forms, to harbour the greatest abundance of indigenous plants and wildlife that means we should embrace them. They can be scaled down and formed in such a manner as to replicate their ancient countryside ancestors but still fit within the context of a garden space. We can also adapt them further, highlighting only their wildlife-harbouring and privacy-creating attributes while doing away with any function for holding in or out mammalian life. Hedgerows are wildlife's pathways and byways from one area to another, and if more people create them, they also become corridors from one garden to another.

Table 3.1. Common Hedgerow Species

Trees & Shrubs	Scramblers & Ramblers	Bulbs & Perennials
Blackthorn (*Prunus spinosa*)	Black bryony (*Tamus communis*)	Bluebell (*Hyacinthoides non-scripta*)
Dogwood (*Cornus sanguinea*)	Bramble (*Rubus fruticosus*)	Foxglove (*Digitalis purpurea*)
Field maple (*Acer campestre*)	Honeysuckle (*Lonicera periclymenum*)	Red campion (*Silene dioica*)
Hawthorn (*Crataegus monogyna*)	Ivy (*Hedera helix*)	Solomon's seal (*Polygonatum multiflorum*)
Hazel (*Corylus avellana*)	Sweet briar (*Rosa rubiginosa*)	Wood anemone (*Anemone nemorosa*)

Hedgerows benefit from regular pruning, which helps them thicken. This should only ever be carried out in the winter, when all the leaves and berries have gone and there are no active bird nests present. Most tree and shrub species do not require cutting every year, however, so rotation pruning is desired to allow the plants to produce flowers and berries on old growth. This could involve cutting only the top one year, one side the next, then the last side the year after, depending mostly on the height and size of hedge that you desire. After five years or more of this schedule, the hedge will have thickened out and a slightly different approach will be essential to maintain the health of the hedgerow.

Bramble. The most common scrambling plant for wild foraging.

As time progresses, the trees and shrubs in the hedgerow will follow their natural path and reach their mature

LAYING A HEDGEROW

This traditional method is associated with creating an impermeable barrier for livestock, but it can be reimagined in the garden setting, not only to provide the regrowth the hedge will need for health and vigour, but also to make a hedge with rustic charm and traditional craft. This type of hedge will perfectly suit situations where space is at a premium and also if a hedge is running through a garden as an aesthetic punctuation or wildlife corridor. In the end, the style of the hedgerow is irrelevant; as long as it provides habitat for wildlife, and privacy and shelter for people, we'll have a linear biodiverse ecosystem instead of an impermeable barrier.

1. Cut the main stems of all the shrubs and trees in the hedgerow almost all the way through their trunks. Then bend these stems - known as pleachers - so that they lay on their sides.
2. Repeat this process throughout the hedgerow with each successive stem cut and laid parallel on top of the previous one, creating an angled (45 degree or less) stack of trees and shrubs.
3. To hold the stack together, use uprights with stakes made from thick hazel poles, set approximately 1–2m apart, and binders - made from smaller hazel stems - woven along the top of the stakes.

An example of a recently laid hedge with the fresh cut of the pleachers still visible.

height. In the case of field maple, or any other native trees you may have placed in the hedgerow, this could be as tall as 15m or more, so you must guide the hedge with a different pruning approach. Species such as hazel and field maple will benefit from being coppiced down to the ground, as this will reinvigorate them within their compacted environment and ensure their healthy survival. The gaps produced are part and parcel of creating a permeable boundary, but there is another method that will create no sizeable gaps and form an entirely new beginning for the hedge: laying (see Laying a Hedgerow on page 96).

I've spent more time than I'd like to remember cutting hedges, with their popularity as robust as their contoured branches. From the tiered topiary of the most established gardens in the land to the ballooning bobs that litter suburban front gardens, I've cut them all. More often than not, these are nothing more than green walls, expected to form regimental uniformity of shape, irrespective of the plants' natural habit. But I have also been lucky enough to plant and nurture many country hedgerows, watching the small whips turn to stronger saplings and eventually weave and jostle into tangled thickets. By the nature of my work, I get closer than many to the inner workings of various hedges, and I can say without compunction that indigenous species hedgerows are alive, whereas mono-hedges in all their grandeur and shape are barren, lifeless places where even the light doesn't want to go.

Weeds in the Wild: A Case of Mistaken Identity

When raising a garden of indigenous plants, the ecological gardener will inevitably come up against that most perennial of anti-labels: weeds. The concept of weeds seems inextricable from gardening and to wrest it apart is no simple task, but we must make the effort, as it leads to some of gardening's most destructive tendencies. Weeding has, unfortunately, become a byword for neatness and order when it should be about selectiveness and restraint. Weeds have historically been defined as a 'plant out of its place' but gardens are ecosystems, dynamic entities undergoing constant change and upheaval, and so a moving mosaic of plants is only its natural course. Ecological gardeners endeavour to shape and guide their gardens, but with knowledge and wisdom that our actions should only promote more diversity, abundance and health. The definition of weeds needs to be revised thus: all indigenous plants are wildflowers and all non-native invasive introductions are weeds. This reverses the conventional gardening perspective that looks at many native plants as weeds and yet will happily promote invasive non-native species to even greater abundance (see table 3.2). Indigenous-species protection is the bedrock of almost all conservation efforts in the wild, but for some reason gardens stand alone as antithetical to this idea. As regenerative gardeners, we can only look to tread a path away from the ideas of the past and hope that this course of action will help the traditional consensus to shift.

Virginia creeper. Easily jumps from garden to wild where vigorous growth takes over.

It can take time to learn how to identify plants not by their flowers and general appearance but by their juvenile leaves, and yet this is the cornerstone of wisdom on which all weeding relies. Take, for example, teasel (*Dipsacus fullonum*); as a biennial it sets seed the year after becoming a seedling and then dies. If you want to continue to have teasel in your garden, you must first allow it to set seed but also nurture the growing rosette of leaves in the first year before the stems or flowers even appear. Knowing how to identify teasel just by its leaves will change the focus of your

Table 3.2. To Weed or Not to Weed?

Invasive non-native species that are readily planted and encouraged.	Indigenous species that are viewed as invasive weeds and removed.
Giant rhubarb (*Gunnera manicata*)	Comfrey (*Symphytum officinale*)
Lady's mantle (*Alchemilla mollis*)	Daisy (*Bellis perennis*)
Spanish bluebell (*Hyacinthoides hispanica*)	Dandelion (*Taraxacum officinale*)
Virginia creeper (*Parthenocissus quinquefolia*)	Teasel (*Dipsacus fullonum*)

gardening method and mindset; some can be removed if they are too dominant, while others can be left for their continued survival.

The key difference between many of the species that you might source and grow for a specific area, like a meadow, and the ones listed in this section is that you don't need to look far for these wildflowers – they will find their own way. That is part of the reason they attained their status as weeds, but their amazing adaptability is to be admired and understood rather than vilified and rejected. Their omnipresence is down to one thing above all: reproduction. Most of the common wildflowers that find their way into our gardens are phenomenal reproducers, and their unmatched ability to survive has led them to flourish while other species have slipped into decline. If we can better understand their ways and means of reproduction, we can not only break down their fearsome façade but also monitor their spread to check any progress that would reduce the biodiversity of the garden.

Plants reproduce in one of two ways: by seed or vegetatively, and many of the most efficient survivalists will do both. Annuals and biennials will always reproduce by seed, while perennials will either use vegetative means only or use both seed and expansion. The annual and biennial wildflowers that we see the most have remarkable tendencies to not only produce huge numbers of seeds but also to make seed that can stay viable in the ground or produce successive seeding offspring throughout the year. These include hairy bittercress (*Cardamine hirsuta*), chickweed (*Stellaria media*), fat hen (*Chenopodium album*), groundsel (*Senecio vulgaris*) and herb robert (*Geranium robertianum*).

Herb robert. Annual or biennial, it is found everywhere but usually in shady spots.

Many in this class are derided as much for their appearance as their invading threat, hence their weed moniker. But the truth is that they, like many annuals, are just pioneers of bare soil. We need to remember that the soil is positively loaded with seed already, and this is only natural, so if you create an entirely unnatural situation, like bare soil, you should expect to see these species and they should be allowed to follow their natural cycle. However, if you follow a method of no-dig mulching and perennial cover, you are unlikely to see these types of species occurring regularly in the garden.

It's only when we reach the vegetative reproducers that we meet plants with designs so cunning that it can take thoughtful due diligence to learn to live alongside them. But live alongside them we must, as almost every one of these perennial pioneer species has a deeply bonded relationship with wildlife that relies on their continued existence to survive.

Bramble (*Rubus fruticosus*). A multitude of wildlife species depend on bramble's ample bounty, from the nectar in its flowers to its fruits in the autumn and the habitat created by its stems. Bramble reproduces by creating new roots from the tips of its arching stems (and by copious seed) but you can treat it just like any other soft fruit. Either grow it in a raised bed along a trained support and prune every year after flowering, or allow it to ramble through a native hedge and cut it back when its growth gets excessive.

Rosebay willowherb. Its alternative name is fireweed, as it is able to colonise land after fire.

Rosebay willowherb (*Chamerion angustifolium*). This indigenous wildflower employs the dual reproduction methods of vegetative expansion and seed dispersal, but as with most perennials, it is mainly the vegetative growth that we need to be fully alert to. Rosebay will spread by rhizomes, which means it will send out roots parallel to its main rootball and send up new stems from the freshly grown rhizomes. All we need to consider is whether we have the space to allow this to happen, and if not, either check this growth perennially through digging some of it out, or place the plant in a barrier bed that won't allow its rhizomes to spread.

Field bindweed (*Convolvulus arvensis*). Possibly the most vilified of all our native species. Not for its pretty white trumpet flowers or its fundamental relationship with many wildlife species, but for its ability to reproduce from just the smallest portion of its root and its habit of climbing and clambering all over other plants and smothering their growth. It also sends down a deep taproot and its seeds have extremely long viability in the soil. All of which adds up to an impressive CV that we should admire for its invention and ingenuity. Bindweed will need to be treated like a climber and given support to grow up, and its roots will need to be barrier-contained, but if we can live alongside it, we can live alongside nature at its rawest and the diversity of the garden will be the greatest benefactor.

Although you might need to restrain some specific species from spreading all over your garden, overall, look to loosen the reins of control over your wandering wildflowers. Allowing areas of the garden to be untouched by intervention is the fastest route to rewilding your space and reaping the benefits of an increase in wildlife biodiversity. These wildflowers already grow anywhere, so letting them take hold in the otherwise neglected spaces between sheds and walls or around the compost heap or log piles is the easiest place to start. The transient nature of these species will always mean that they spread beyond their place of origin, but this is the way of a dynamic garden and fighting against it can sometimes be futile. Dedicated areas like nettle beds, raised barrier beds, bug beds or containers are certainly the most effective ways to rein in the most rogue and rampant species and are well worth the time and effort. However, these strategies need only be employed for a small number out of the great expanse of wildflowers, and living alongside them and gaining knowledge of their habits will always be among the most rewarding of gardening endeavours.

Wild Abundance

When nature is allowed more freedom to pass through the garden, we will inevitably accumulate more plants and more diversity. However, the need to create and reproduce more ourselves will remain, if only because we are gardeners and raising plants is what we do best, whether by collecting a plant's seeds, dividing its roots or taking cuttings from its leaves or stems. We will cause no undue harm to the plant by doing this, and in the case of dividing we can invigorate the host plant to better health and longer life. I believe this is a central pillar of ecological gardening – to propagate more life from what we already have, even if that means a surplus to our needs. Through the skill of the gardener, one plant has the ability to become tens if not hundreds more, and each one of those offspring tens or hundreds more again.

Different plant species require different methods. For instance, annuals and biennials can only be propagated by seed, while trees and shrubs will have the greatest chance of success if reproduced from a cutting. We can propagate herbaceous perennial plants every which way, but generally the division method suits them best. Propagating is trial and error, and I like to try as many different methods as I think the plant will allow and see which practice produces the best outcome. It's always better to produce more than you need and imagine your success rate far lower than to propagate an insignificant amount and expect all to germinate or strike every time.

Seeds

Collecting seed is easily one of my favourite gardening tasks, as I love to look at the minuscule seeds in my hand and wonder how nature transforms from just one seed into the mature plant standing before me.

Once you have collected the seed, you must decide where you're most likely to sow it next, as this will influence how the seed should be kept. In nature, the seed will lie on the ground from the moment it's released to wait for germination in the spring or maybe even many years hence, until the right conditions arise. We can replicate this by simply broadcasting (scattering) the seed, still attached to the stems, where we wish them to grow without even needing to collect the seed at all – this works well for biennials like foxgloves. Or, we can collect the seed in the way described in Collecting Seed and choose to either broadcast it in the autumn or hold some back for seeding in the spring. If so, some knowledge of seed dormancy is required.

Many wild species produce seeds that depend on particular weather patterns and events in order to germinate at the optimum time for their survival. These include frosts, heavy rains, low light levels and fluctuating temperatures. Dormancy is an evolutionary adaptation that prevents seeds from germinating if these conditions are not met, so if we store seeds inside and try to sow in the spring, they won't produce plants. Therefore we must use a process called stratification in order to replicate this wild dormancy period. The best way to achieve this is to place the seed

COLLECTING SEED

Collecting seed isn't as easy as it seems; it takes timing, experience and judgement to collect at the perfect stage of ripeness.

1. Once the plant has flowered, watch for the stem and the seed-head to brown and ripen.
2. Collect on a dry day. Cut the whole stem off with the seedhead attached and place in a paper bag or something similar.
3. Take the stem inside and carefully extract the seeds on a table with white paper spread out on top – seeds can be small so the paper will help you detect them when they fall.
4. Once you have released all the seeds, collect them up and place them in small envelopes or sachets, then label them.

Collecting seed is one of the great pleasures of the gardening year.

Table 3.3. Stratification and Scarification

Species Requiring Seed Stratification	Species Requiring Seed Scarification
Cowslip (*Primula veris*)	Bird's foot trefoil (*Lotus corniculatus*)
Monk's hood (*Aconitum napellus*)	Hare's foot clover (*Trifolium arvense*)
Pasqueflower (*Pulsatilla vulgaris*)	Meadow cranesbill (*Geranium pratense*)
Primrose (*Primula vulgaris*)	Meadow vetchling (*Lathyrus pratensis*)
Yellow rattle (*Rhinanthus minor*)	Salad burnet (*Sanguisorba minor*)

within a slightly damp medium, such as very gritty compost or damp paper, seal it in a jar and place it in the fridge or freezer for a period of time – generally 6–12 weeks. Some species require further manipulation, called scarification (see table 3.3). This involves breaking the hard outer husk of the seed by either cutting or crushing, then allowing the internal embryo to break out and germinate. Finally, if we are only seeking to store the seed, say until sowing the following autumn, then it's best to keep the seed completely dry in a sealed jar in the fridge or a dark cupboard.

If the seed has been through the aforementioned processes, then come late winter or early spring it can be sown into pots or trays in a greenhouse or on a windowsill. If the seeds are small, then broadcasting them into trays is the most effective. When they break through and form seedlings, you can thin out around half to allow the others to grow strong and healthy. As these seedlings grow bigger, tease out and place them into a small pot to grow on further. If the seeds are large enough to handle individually, place them one by one into small pots and grow on until they can be transplanted into larger pots or out in the ground. Remember to use low-nutrient soil in the beginning and move progressively up into higher nutrients as the plants progress – for example, from leafmould and sand to compost and loam.

Cuttings

Cutting is the art of taking a side shoot from the main stem of the plant you wish to propagate and using that cutting to grow a whole new one (with some plants, you can also do this with the

TAKING CUTTINGS

Propagating through cuttings is an entirely different animal from collecting seeds, and requires a certain level of skill and knowledge to get right.

1. With a sharp knife, cut 10–12 non-flowering, fresh-growth side shoots (branches off the main stem), at 100–150mm long, from the plant you wish to propagate. The best side shoots are usually tucked away in the middle of the plant, as in the case of common knapweed (*Centaurea nigra*, figure 1).
2. At a table or bench, cut all but the top 2–4 leaves from each cutting all the way back to the node at the base of the leaf on the stem (figure 2).
3. Fill a pot with very gritty, free-draining compost. Use a dibber – a pencil will work fine – to make holes for the cuttings and insert the cuttings around the edge of the pot (figure 3).
4. Mist-spray the cuttings thoroughly and at regular intervals over the coming weeks, and keep the pot out of direct sunlight.

1

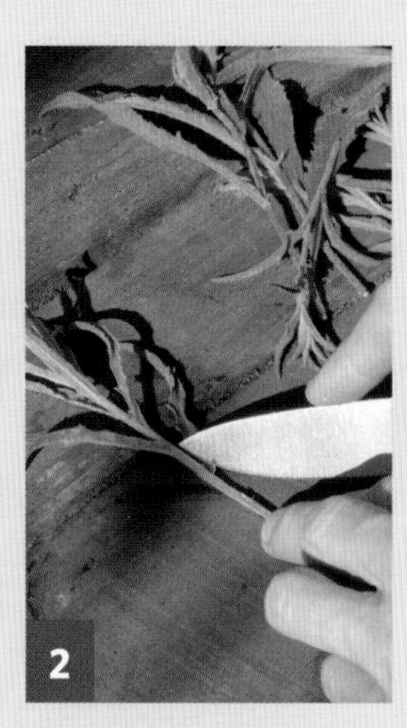
2

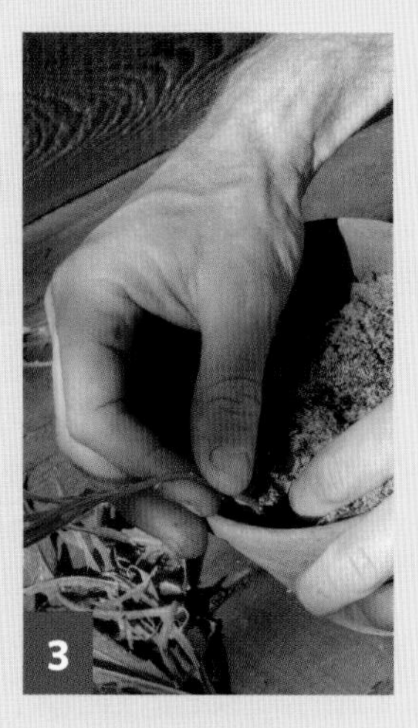
3

root). You can take cuttings from spring to autumn, depending on the species; stems at each stage of the year will be different, and require their own methods of propagation.

A stem cut in spring will be soft and floppy; if it's cut in autumn, it will be harder and straightened. This is important to understand, as once it's cut, your job is to encourage this stem to grow roots from only the small leaves it has left at the top. At this stage, the soil you use to pot the plant is incidental; your only chance of success is to keep the stem and the leaf in a constant supply of moisture until the roots begin to grow at the bottom of the stem. With a floppy spring stem this can happen a lot quicker, but so can the rate of dehydration and death of the stem. If the stem is hard and cut in the autumn, the progress of the root's formation will be slow but the chances of it drying out much reduced. Mist-spray the freshly cut stems as much as you can, twice a day at the very least. As soon as you notice new leaves emerging, you'll know it has produced roots and you can then transplant into a bigger pot and grow it on just like any other seedling or juvenile plant.

Division

Reproducing plants through division is an ancient propagation method, usually best carried out in the autumn, but it can be achieved in spring and summer too. It simply involves digging up a host plant and dividing this plant into many sections that will become new plants in themselves.

Divided plants can either be planted out in the garden straightaway or kept aside, to be given to family and friends or planted at a later date. This method is probably one of the most crudely effective and easiest to get results from. But it's not my favoured method, as digging damages the fine fungi filaments and microorganisms in the soil. If there are other methods available, try these first.

I have always sought to propagate species already present in the garden, possibly from my days working in public gardens, where the 'show' must always go on. But it instilled in me the value of such practices as a way to bottle nature's amazing ability

DIVIDING PLANTS

For the division method, the plant to choose will usually be a perennial - such as a primrose or ox-eye daisy - that doesn't look as healthy as it once was, or needs its progress checked, or is simply one you would like more of.

1. Dig the host plant out of the ground or take it out of its pot and shake off excess soil - the key is to remove *all* of the soil (figure 1), as shown here with purple loosestrife (*Lythrum salicaria*), a native perennial.
2. Tease apart separate stems and their roots - or, if the plant is large and densely rooted, cut or saw the rootball into sections. Most perennials have multiple individual stems with their own roots (figure 2).
3. Pot up each new plant into its own pot in home-made compost, water thoroughly and set outside. Newly divided plants should begin to grow anew rapidly (figure 3).

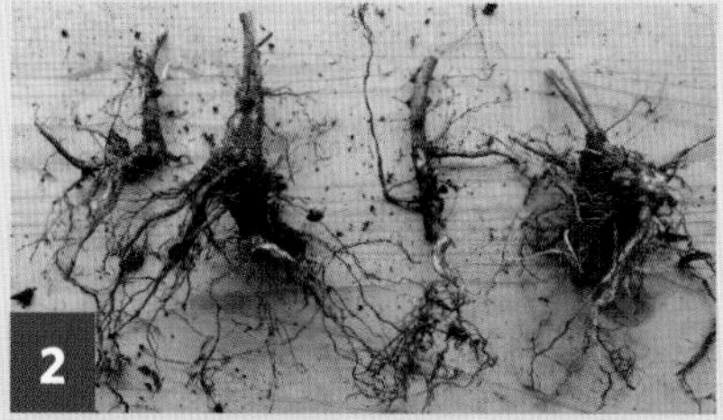

to reproduce, and to foster abundance from paucity. I am always astonished at how, from just a few tiny seeds, many decades of life can be raised; and the more we grow our gardens from these humble beginnings, the less we will need to rely on buying anything new.

Built to Last: Native Trees

As owners of a parcel of natural space, one of the very best things we can do to readdress the current imbalance in our fragile environment is to plant the acorn of an oak (*Quercus robur*), stand back and let it grow. Our lifetimes will be gone many times over by the time it fully fulfils its majestic arc of significance, but in giving it life we may never do more for the natural world. Everything about trees points towards restitution and regeneration. From their ability to draw down and store gargantuan amounts of CO_2 out of the air to the sanctuary and sustenance they provide for wildlife, trees are like lighthouses guiding us away from our most destructive of impulses. They are also arguably the most beautiful of all the plants in the garden, with their steadfast trunks and radiating branches providing the structure and bones of a garden's shape, while their leaves, flowers, fruits, nuts and seeds give colour, shade, fragrance and food. But if their appearance above ground is rightly admired, below ground their function is of even greater significance, as their roots regulate the water content in the soil and stabilise it against the forces of erosion and desertification. Once they die, their significance to the soil continues as they return their mass of organic matter. Trees are the pillars of the natural world and without them, everything falls down, so growing them needs to be an essential part of our garden planting.

In the average-sized garden, sadly the oak tree will eventually envelop much of the space, and although it will take a long time to do so, it may not be the most practical of acquisitions. However, if you have a wealth of space then, by all means, plant an oak, beech (*Fagus sylvatica*), ash (*Fraxinus excelsior*) or small-leaved lime (*Tilia cordata*). But for those with a more average amount of

The oldest oak in the UK is said to be over a thousand years old.

space, smaller, more diminutive species may be called for. Certain species can be pollarded, whereby a trunk or stem is cut regularly at a height to promote the growth of a dense head of foliage and new branches. Others can be coppiced, which is similar to pollarding, except that the cut is made at the ground. Both techniques allow you to grow trees within a smaller space and will even aid the growth of some species, such as crack willow (*Salix fragilis*), but we must remember that cutting will forever change their natural shape and form, and may not always be desirable. Thankfully, there are many other indigenous species that will grow to a proportioned size for any garden space, and here are some recommendations.

Crab apple (*Malus sylvestris*). The original and the best. It would take more than this entire book to describe the complex wealth of history that has sprung from cultivating apples from this original source, so I won't. Instead, just plant it to admire its beauty. However, don't expect apples, at least as we now know them. Crab apples are small, hard and sour, and much better in jellies and jams. These traits are also an indicator of an original species, not a cultivated one, as it follows logical sense that we would have cultivated towards an edible fruit. What has been forgotten along the way, however, is that the fruits are more than appetising to wildlife, and with more and more crab apple trees disappearing at the expense of cultivated varieties, planting one will go a long way towards rewilding your garden.

Hawthorn in full bloom.

Common hawthorn (*Crataegus monogyna*). The May tree is just too spectacular and lettered in folkloric veneration for use solely in the hedgerows, where it's abundant. To see flowers in May on a fully mature tree is a wonderful sight, and the tree is more than able to hold a candle to any other flowering tree, from anywhere in the world, of similar size. Its colourful history includes being the keystone plant in the pagan May Day celebrations that heralded the end of winter and beginning of summer. And also, by being the ancestor of the Glastonbury thorn (*Crataegus monogyna* 'Biflora'), the May tree is thought to have been brought to England by Joseph of Arimathea – Jesus' great-uncle. But I'll leave you to decide between fable and fact.

Hazel (*Corylus avellana*). You cannot have a garden of indigenous British plants and not have a hazel tree. Hazel, like willow, is so adaptable that it can be left to grow to maturity or be grown as part of a hedgerow or as a stand-alone coppice. It's important to state that coppicing will do no damage to the tree – in fact, it will lengthen its life considerably – and be of no less benefit to wildlife. Coppicing was originally carried out to harvest the useful stems produced from such practice, but the coppice can be done completely at your own discretion. We want to encourage biodiversity and provide for wildlife, so the coppicing will not be done every year but every five years or ten years. Or you may only take a few stems out one year and a few more the next, then wait another few years before doing

A recently pollarded willow showing whippy fresh stems of new growth.

more again. Whichever way, the hazel will survive, thrive and provide wonderful natural materials along the way.

Crack willow (*Salix fragilis*), pollarded. This tree will need to grow in an area that is constantly wet, like the edge of a pond, river or bog. If you can provide that, then you will also need to pollard the tree to about 1–2m approximately once every three years. This encourages branches to shoot up from the cut like billowing grasses. You'll restrict the tree from ever reaching its full maturity, but the beauty of the crack willow is its natural inclination to do this anyway due to its branches 'cracking' apart and collapsing under their own weight. They then send out new stems and branches from the open wound, which pollarding imitates. So you'll be doing nothing out of the ordinary for the tree and be able to grow it within a garden and take pleasure in its gnarled, olive-like beauty.

Tips for Planting a Tree

When planting a tree, there are a few key considerations to ensure it has the best chance of beginning life in full health and vigour.

1. Only plant a tree in late autumn, winter or early spring, when the water table is high and the rains are most frequent.
2. Dig a hole wide enough that it allows room for the roots to grow into loose soil and don't compact the soil when backfilling the hole.
3. Don't add compost into the hole; instead, add a thick layer of compost around the base of the tree once planted.

Tree Holistics

Planting and maintaining trees relies on a suitable amount of knowledge and practice at the beginning that will eventually lead to the most self-sustaining of plants as they begin to grow to maturity.

Once the tree is established, however, you should have no problem at all with its growth and health, as indigenous trees are far superior in durability and toughness to all other species – they are exactly where they're supposed to be. They've also formed bonds and complex interdependent relationships with the native wildlife that run so deep as to be immeasurably significant. With the right trees, our gardens can be the engines of the regeneration of these bonds, forged over many thousands of years, allies in the fight against the volatility of our precious climate. If we look beyond the exotic and unusual and back into the heart of this country's indigenous flora, we'll be amazed at the treasures that await us.

Freshwater in our rivers becomes more polluted as our use of the mains system increases.

CHAPTER 4

Water

Whether moved by rivers, held in lakes and aquifers, or suspended in giant ice sheets, freshwater is a precious resource to revere and respect, not least of all because we – and the plants in our gardens – rely on it for survival. Some parts of the world possess water in abundance, while others have evolved strategies to conserve and make the most of what little rain they receive. Here in the UK, winters in particular can bring torrential downpours, causing destructive floods from bursting rivers and saturated ground. Why then, we must ask, do we still need to irrigate our gardens with freshwater from a tap?

The truth is that most of the rain we receive drains away into engineered water systems that funnel it out to sea, leaving a parched and thirsty land only a few weeks after the late-winter downpour has ended. To watch great waves of freshwater wash over our gardens through the winter months, only to wonder where it's all gone in the summer months and reach for the mains tap is a sort of folly or madness. So to capture, harness and utilise this free and abundant elemental force would seem to be the most logical and simplistic path to follow. Not only does it mean bypassing the great energy apparatus of the mains drainage system, but it also allows us to protect and regenerate the ailing ecology of our river systems and replenish our underground aquifers. While the practice is as yet uncommon, it is possible to water our plants only by natural means, and in doing so further maintain and enrich the health and vitality of our soils. Water is a fundamental element in the garden ecosystem and how we source, use and reuse this resource is of vital importance.

There are many ways to turn this aspiration into reality. Whether you re-engineer your drains and gutters to make better use of the rain that falls on your roof, or carefully manipulate the land to cradle the rains instead of allowing them to wash away, you can create an entirely new framework for water in the garden. You can even recycle and reclaim the freshwater you use in your home, allowing it to travel back to the place from which it was plundered.

As gardeners, it's always been easy to admire the helping hand Mother Nature provides when the rain comes lashing down, but how it moves, gathers, and acts underground – percolating through the soil or rising up from the water table below – can be as mysterious as the microbes in the soil. Instead, we can bring it out of this subterranean underworld and into the visual realm, where we can admire it and put it to more use. Ponds, swales, ditches and barrels can collect and suspend water, while rills, gullies and rain chains can play with its shape and form as it traverses across the garden. Harvesting and reusing freshwater gives you a reserve of water to help your garden endure drought, while slowing down and collecting water in ponds and other features can help mitigate floods and erosion. By learning to work with water and understanding its nature, your garden can be reliant on nothing more than the rain that falls from the sky or the water otherwise thrown out as waste.

Rainwater Harvesting

There are both simple and sophisticated methods of rainwater harvesting. Enlightening ourselves about its various forms can be an empowering endeavour that brings us closer to the ingenuity of the earliest civilisations, which mastered the art of harvesting from the rain. The Minoans, for example, built highly sophisticated networks of drains, cisterns and sediment tanks that would harvest and process rainwater to use throughout their towns and cities. And the Romans, the veritable masters of water engineering, constructed enormous aqueducts that funnelled the water that fell on the mountains into the metropolitan hearts of their cities. But perhaps there is no greater example of rainwater harvesting than the truly monumental Basilica Cistern in Istanbul, capable of holding a mind-boggling 80 million litres of rainwater. Although the Basilica is now dry, its use superseded by the mains water system, we can take heart from the awesome possibilities that rainwater harvesting possesses. Happily, as we are only concerned with the water needs of our gardens, our task is less Herculean than that of our early ancestors, though it's no less vital to undertake, and of an equally imperative and prescient need.

Harvesting Potential and Need

How much water does it take to sustain a garden over the course of a year? There are so many factors at play that it's difficult to find an exact answer to this question, but we can begin by looking at how much mains freshwater we currently use in the garden.

Households in the UK use, on average, 350l of freshwater every day. Conservative estimates say 10 per cent, or 35l, of this is used in the garden. That puts an estimate for the entire year at 12,775l. If we want be more conservative and discount all non-gardening pursuits, like washing a car, we could say that we use 10,000l of freshwater in the garden every year. Let's then further say that through being more conscientious about our water use, and how, what and where we plant, we can reduce this heavy reliance on water and cut this figure by half to 5,000l. This means that if the average gardener can harvest 5,000l of rainwater per year, they will have extinguished the need for mains freshwater use in the garden. So, can this be done, and how?

The most common harvester is the humble plastic water butt, the standard size of which holds 150l. Assuming one household has two of these attached to their gutters, and that over the course of a year they are emptied out and filled up again eight times, this still only equates to 2,400l – way short of the 5,000l target. In my experience, gardeners with this set-up also come up against the perennial problem of praying for rain in the summer months, to fill the water butt back up, when it's least likely to rain! This makes the figure of 2,400l wildly optimistic and confirms the need to store more water in the rainy months, ready for use through the dry months.

Now let's consider how much water we *could* harvest from our roofs and through our gutters. In the UK, the average annual rainfall is approximately 850mm. Even if we base our estimate on the driest area in the whole country, which receives 450mm of precipitation per year, and the smallest average roof size of 50m^2, the resulting yearly rainwater harvest would be approximately 22,500l. It's hard to imagine how we would capture all this, but it shows the quantity is there to be harvested. So if we were to double the size of the harvester to 300l (the size of a Barrique barrel), double the quantity of harvesters attached to the gutters to four, and reduce the amount of refills through the year from eight to four, we would reach the figure of 4,800l. It's still short of the 5,000l target but a vast uptake in harvesting potential, and certainly a more realistic expectation of capture through the dry season.

There's clearly a glaring divide between how much most gardeners attempt to harvest and how much can actually be harvested. Once you understand your garden's needs, and accept that you should – and can – harvest in much greater volumes, then together with other water-saving techniques, you can easily forgo all your mains water use. And, more importantly, you will see no reduction in your ability to use water to grow seeds, water pots, and shepherd the garden through even the driest of summer months.

Water and gardens go hand in glove, and I've been at the coalface of this relationship for most of my working life. I've implemented harvesters and ponds wherever and whenever I've been able to, but I've also, as a younger man working for others, installed lawn irrigation systems and run large standpipe sprinklers. The problem is, here in the UK, for nine months of the year the issue is usually an overabundance of water, and most thoughts and efforts are driven to draining the water away. Then the summer months come along with ever more frequent droughts, and since nothing has been done to prepare for this eventuality, gardeners are hopelessly reliant on the mains to save their rapidly ailing plants.

Water retention may not seem like a necessity the way it is in the driest corners of the globe, where gardeners can only dream of harvesting the amount we can, but it's still important, just for different reasons. In my garden, I have complete control over what I can and can't do, but convincing other garden owners and the established horticultural industry takes more diplomacy. Huge rainwater tanks are a common sight in the dry rural regions of Australia and the US, and if we see more rainwater harvesters of appropriate size in the public and private gardens of the UK, the more successful the persuasion will become.

Harvesting Methods

Collecting rainwater from our roofs can seem to be such a perfect marriage of practicality and convenience that it's easy to fool ourselves into believing that's what the gutters and drains were originally put there for. Of course, the opposite is true; their primary purpose is to funnel the water away from the house and residential area as quickly as possible. Water travels from most gutters into the storm drain system that will link off with the sewage from the house and eventually end up in the sea, bypassing all the natural underground reservoirs and river systems it would have naturally replenished. However, since this system is firmly in place, the best we can do is repurpose its efficiency to both our advantage and that of the natural world. There is nothing simpler than fixing a rainwater harvester to our house drains and gutters, and no better way to harvest water in large volumes.

Of course, houses aren't the only structures that have roofs, gutters and drains. Greenhouses, potting sheds, barns, summerhouses, woodsheds and anything else with a roof have the ability to harvest rain, and you can install gutters, drains and collectors on them if they don't have them already.

FITTING A WHISKY-BARREL RAIN COLLECTOR

Using a reclaimed wooden whisky barrel or a Barrique barrel is one of the very best ways to collect and store rainwater. They look fantastic in the garden, adding rustic style and charm, and upcycle materials whose commercial use has ended.

1. Acquire a used whisky or wine barrel, either direct from the whisky/wine producer or through a supplier online (the standard barrel volume is approximately 300l).
2. Fit a metal tap to the bottom (figure 1) and cut a hole in the top for the downpipe.
3. Remove the existing downpipe and place the barrel on a plinth of bricks underneath the downpipe point of the gutter.
4. Cut the pipe to the height of the barrel, then fit an angle bracket from the downpipe into the harvester (figure 2).
5. Consider where the overflow will go once the harvester is full - preferably into a rill or gully to carry the water to a holding pond or swale (figure 3).

Even though it's clear that your roof can provide more than enough water to meet your garden's needs, that doesn't mean it's the only way to collect rain. Stand-alone harvesters can be employed too. Essentially, these are just collection vessels with a funnel or saucer on top – imagine an upside-down umbrella in a barrel and you're not far from the reality. When you remember that $1m^2$ of surface area (the umbrella) receiving 500mm of annual rainfall will collect 500l over the course of a year, then you can see the power of harvesting rain in a relatively small amount of space. A harvester can even be as simple as a tarp strung over a barrel, or an integrated canvas atop a large cistern. Stand-alone harvesters also have the not-insignificant advantage of being unencumbered by attachment to a building with roofs and gutters, meaning you have the freedom to place and move them anywhere in the garden, wherever the water is needed most – say by a productive bed or holding area. You can also place them at any height, such as on a flat roof, which will enable you to better harness the force of gravity when releasing water from the vessel.

Although they don't technically harvest rain, dew and fog collectors can collect a surprising amount of water from just the air. Working not too dissimilarly from dehumidifiers, they attract water vapour onto their surfaces, which will condense when the weather changes and turn to dew or water that can then be harvested. They usually follow the same design principle as the stand-alone collectors, with a funnel or saucer positioned above a collection vessel. The crucial difference is that a fog collector's saucer needs to be made of mesh fabric or a material that will become cooler than the ambient temperature, like metal. These work best in arid climates or high above the treeline in the mountains, but that shouldn't prevent them from being utilised during the summer months in a more temperate climate.

Practical Storage

If you are looking to implement the barrel rainwater collection system, but have a garden larger than average or consider your water needs to be greater than most (say for a large veg garden), then you may need to consider linking these barrels into sets for added capacity or look into larger vessels like cisterns and tanks. These sets can be lifted off the ground and mounted on the wall, both to allow extra space and to apply more gravity pressure and therefore power to the water when it is released through a hose. If possible, using gravity pressure to water your garden should always be the preferred option. If space allows, siting your collection tank on a flat roof below a pitched roof with gutters and drains will make a sizeable difference in your ability to use water under pressure in the garden. If you are even luckier still and have land beyond that is also higher in elevation than the garden, then building a roof above a static holding tank at the highest point will even more effectively enable that water to flow down to the garden under pressure. However, most of us don't have that luxury, so if groups of barrel sets don't suit your garden scenario, then the next option is a singular large cistern or tank that can hold from 1,000–10,000l of rainwater. If you have the right amount of space, you could accommodate a cistern that large, much like you would a large oil tank. However, if space is at a premium or you don't

Large rainwater harvester made from corrugated steel. Photo by JCL Garcia/Shutterstock.

want it to be the dominant feature in your garden, then the last option is to place the tank underground. Most underground tanks have a capacity of up to 7,500l, but this is the equivalent of an underground well and if you have a well, you need a pump.

I believe that if we are harvesting rainwater, it's not only because it's an abundant and free resource, but also because we want to reduce the energy needed to pump water to and from our houses and gardens. If we then use electricity to power pumps in underground harvest tanks, I think we will have missed the point. This doesn't mean that we can't use pumps – they just need to be the hand pumps of old, or powered only by solar energy. My preference is for the Victorian-style hand pump that will do the job just as well while also being an attractive addition to the garden.

Aesthetics and Materials

It only takes a cursory glance online to see that almost every one of the storage butts, barrels or tanks for sale are made of plastic, and each one uniformly militaristic in design. Looking at them in the garden is about as pleasing as looking at a waste bin or an oil tank, and it's easy to see why most gardeners avoid them altogether or hide them underground. I have always felt that every object in the garden needs to be in harmony with its surroundings, and that's especially true for something as big as a 5,000l tank. Harvesters can and should be beautiful and integral to the design and aesthetic of the garden; and this can be achieved by rejecting the new and looking instead to recycling and repurposing the old and the used.

You can make small collectors from reclaimed oak whisky barrels, rustic steel oil drums, porcelain baths, or even old copper water heaters. Agricultural steel or stone drinking troughs can also be adapted to work well as long and low harvesters in tight spaces. Large, decorative terracotta and stone pots can work too, with the pots almost as distinctive for their attractiveness as for their purpose. If you're looking for a tank that can hold more than 1,000l, you might try to source from a farm and repurpose a small, corrugated steel grain silo or even a small water tower. With a keen eye and perseverance, you might even be able to pick up an old cistern designed specifically for rainwater harvesting. In the end, collecting rainwater doesn't have to be just functional; it can be distinctive and aesthetically pleasing too.

Using reclaimed and recycled materials in this regard are, as we've seen, far easier for smaller-sized vessels than medium-to-large ones. So given that there probably aren't enough used silos, water towers or cisterns to go around, this leads to the question: How do we harvest rain on a large scale without resorting to store-bought containers that are likely made from non-renewable resources like plastic? Keeping ecological considerations first and foremost, we can only conclude to make one ourselves. This way, the only energy required will be our own, and we can choose construction materials not only for their practicality but also for their effects on the environment now and in the future.

One common form of self-made harvester is the rainwater jar. Rainwater jars are large vessels made of mud/cob mortar and finished with a fine clay/lime mixture – capable of holding around 1,000l of rainwater. They are often shaped like large antique olive jars, with a bulbous top that can flatten out to accommodate a small lid. The shape means that you can store more water with less height, and without taking up too much footprint at the bottom. As with most rain harvesters, jars are usually raised on platforms, for ease of access to the tap and to add gravitational pressure. The jar will then be linked to a roof with a gutter and drain system, or be positioned under a stand-alone canvas.

Although making your own jar may seem like it requires a considerable investment of time and labour, you don't have to make it yourself and in fact you shouldn't – many hands make light work after all.

From my experience, the attractiveness of the vessel can be the deciding factor for many people when considering a rainwater harvester for their garden. It doesn't matter if it's a reclaimed whisky barrel or metal farmyard trough, just as long as it's not the feeble plastic butt. This has persuaded me of the need for aesthetical design to go hand in hand with ecological considerations and

Tranquil Effects

If water is being harvested and requires no pumps, then I see no reason why its movement cannot be refined for aesthetic effect. Utilising water in the garden as a visual detail – for its calming, reflective and almost hypnotic effects – is nothing new, and has been practised throughout the ages. Japanese rain chains, for example, have long been employed as decorative and meditative alternatives to the downpipe drain. They consist of links or cups made of copper or steel, laced together in vertical descent from the gutter to the barrel. The barrel, pot or trough used to collect the water can then be left without a lid and allowed to overflow, cascading down the sides and collecting in a basin below that will feed out to a rill or gully. These channels will carry the water to a pond or swale that will feed another pond, and so on. The rills and gullies will ribbon through the garden like the veins through a body, becoming part of the line and form of the garden, and allowing both the gardener and any visitors a chance to admire the movement of water as it journeys from place to place.

Through all my years of working and being in gardens, I have always felt that the space comes alive with the sound of moving water. Even still bodies of water like ponds are never sedentary if they abound with life. The variations with which we can manipulate and play with water are almost endless, and the more water we harvest, the more it can be moved and artfully displayed. If we are harvesting enough water to supply the garden's needs, then as long as we are careful not to allow for wasteful evaporation – keep pools of water deep and shallow water moving – we should experiment with water more.

In Japan, rain chains have been used for their meditative effect for centuries. Photo by fusaromike/iStock.

A collection of handmade rainwater harvesting jars. Photo by Vitawin Choomanee/123RF.

adaptations. Although we can no longer ignore the ecological need for change, if this is married with elegant design and material style, we have the perfect recipe for wholesale changes on a scale that will make a real difference.

Holding Back the Flood

Capturing and storing rainwater is important, but it is only one facet of how you can use water holistically in the garden. Rainwater falls throughout the garden, and the water tables below the surface rise and fall with the changing climate and seasons. How this water moves, changes and percolates through the garden is important to observe and understand. Some questions to ask:

- Does your garden have any wells or natural springs?
- Does a watercourse or river run through the garden, or is there evidence that it may have done so in the past?

- Is your garden on a hill or at the bottom of a valley?
- Is it located between exposed fields and open country, or sheltered in between many houses and buildings?

Gardens exist in a seemingly infinite variety of situations, each with its own distinct ways that water interacts with and flows within. Over time, it's worth studying how the seasons unfold and whether your garden floods during heavy rains or drains rapidly and dries out in the summer. Take note of existing structures like terraces or walls that unnaturally pool water and notice if the soil stays wet long after the heavy rains have disappeared. A depth of knowledge of how water functions in the garden is one of the most invaluable tools in your gardening kit.

Earthworks

Once you have a good grasp of how water moves within your garden, you can consider whether and how to manipulate the landscape to harness its power. The simple way to look at it is to say your aim is to slow the water down – the slower it flows or moves across the garden, the greater the volume you can use to grow plants and the more the soil can absorb. Consider two hypothetical gardens at opposite ends of the spectrum: one, situated on a sharp slope or hill, and the other in lowlands that perennially flood in the winter rains. Water falling on the sloped garden will run off the surface too fast, eroding the hillside and carrying nutrients with it, and so the garden would benefit from a network of ponds and ditches that slow the flow down. The lowland garden, on the other hand, would possibly need to dry out in the summer months or risk becoming a bog. The topography of most gardens probably falls somewhere in between these two extremes, with areas of both moderate run-off and pooling, but there can be little doubt that holding back a body of water from the winter rains to sustain the garden through the dry summer months will always be of benefit.

It's important to understand the subtle differences between distinct types of water catchment and containment that you can implement in the garden. A pond, for example, will need sustaining, with natural fluctuations at a continuous state of water volume; its primary purpose is to store water that can be used again when it's needed. It can have secondary purposes – to attract wildlife, as a storm overflow and as an aesthetic feature – but the main reason to build a pond will always be for storing water. A shallow ditch or swale, however, is principally used to irrigate the area adjacent to the swale – usually hungry or fast-growing plants like fruit trees or coppice trees. Swales and ditches slow the flow of water down so that it has a chance to percolate back into the soil. In contrast to ponds, these may be completely dry for long stretches of the year. But both features are at their most effective when used in conjunction, with swales overflowing into the ponds and ponds overflowing into swales. Imagine a network of swales and ponds crisscrossing along the contours of a slope, and you can begin to grasp how much water can be harvested and absorbed in an area that it would normally rapidly drain away from.

Ponds

I have never been in a garden that was not enhanced by the presence of a pond, and when linked by pipes, rills, gullies and swales to other ponds, the whole network becomes the beating heart of the garden. Ponds can be fed from the overflow of rainwater collectors, by a natural spring or watercourse, or just by the rain that falls. If possible, they should be 1.5m deep with steep sides and sealed with clay, although even the smallest garden can accommodate a pond, with a depth of 50cm and an area of $1m^2$ possibly the smallest you can go. However, the deeper the pond, the cooler the water temperature will be, minimising the amount of water you'll lose to evaporation. Lining your pond with clay will help prevent the water from percolating out. In heavy rains, your pond will undoubtedly overflow, so you will also need to consider where the overflow goes. If the pond is the last stop at the end of the network of water bodies through which water flows in your garden, then you can allow for a soak-away at its overflow point. A soak-away is a pit dug at a depth of approximately 1.5m and backfilled with rocks and gravel, acting as a gateway drain for the water to percolate back into underground aquifers. If it's the first stop, you can direct the overflow out to a swale or to another pond. It's an obvious but pertinent point that if our ponds and rainwater harvesters are full, then additional rainwater will have nowhere to go and will just spill over and race away from the garden as fast as it ever did before. That is why we use swales and ditches – to connect catchment areas and vessels, and to slow water down so more can naturally percolate into the soil.

A natural pond can provide both a water source for your garden and a habitat for wildlife.

Swales

A swale is essentially a ditch dug along the contour lines of a downhill incline, with a berm or mound – usually constructed from the earth removed to create the ditch – raised on its downhill side. On a hillside or incline devoid of thick vegetation (i.e. trees and shrubs) this swale will decelerate the flow of rainwater. The ditch should be dug as level with the land as possible; the goal is not to alleviate or drain the ground of water but to retain the water so it can move more slowly through the soil. In heavy rains, the ditch will fill up and percolate slowly into the berm and then on through the soil. These berms can be planted with thirsty fruit trees or coppice crops of willow (*Salix* spp.) or hazel (*Corylus avellana*), with the ditches providing the plants with a steadier supply of water that can sustain them through periods of drought.

Swales can act as overflow valves for the permanent bodies of water (ponds) in heavy rains or as independent catchments separate from the main system. They can be large enough to

At the bottom of a swale with the last remnants of rain slowly seeping into the berm.

support trees, or smaller constructions planted with evergreen perennials. Regardless, the berms will always be most effective against erosion with constant plant cover and strong root formation to aid them during heavy rain in the winter months.

If you construct large enough berms, they can even be designed to widen out and become paths through the garden during dry periods, making them aesthetic landscaping features in addition to functional ones.

Rain Gardens and Retention

Of course, not everyone has the space to dig a pond, or slopes that could lend themselves to swales. Maybe your garden has lots of impermeable hard standings like concrete patios and driveways, or drains and gutters that just cannot link to any collection systems. If that's the case, it's advantageous to divert the water – which will more than likely be channelled to a storm drain – towards a rain garden. Rain gardens are dedicated areas of planting that absorb rainwater run-off, much like a soak-away, but they use this water to grow plants before sending it into the subsoil.

The same ecological wisdom that advocates slowing down the movement of rainwater – using swales, ponds and rain gardens to bypass using the engineered drain system – applies to strategies for aiding its absorption back into the soil. Soils that are left bare and exposed are far more likely to allow water to run off their surface than soils covered in mulch. Mulch absorbs the water and allows it to percolate slowly through its organic mass, eventually providing a far greater volume of water to the soil below and thereby the roots of the plants. In periods of drought or drying winds, mulch locks in moisture present in the soils and protects it from evaporation on its surface. Any organic sheet mulch will do this to some degree, but thick layers of compost will not only be the most beneficial to the soil but also the most similar to natural processes. Keep the mulch permanently over your soils by applying it in both the spring and autumn, and you can also follow a cut-and-drop method during the summer. Cut-and-drop mulching involves leaving the material left from pruning or cutting

STARTING A RAIN GARDEN

What you choose to plant will vary, dependent on your garden's climate. Gardeners working in dry, arid climates often use rain gardens to grow a variety of plants, while those in wet, temperate climates will necessarily employ a more wet-tolerant bog planting style described here.

1. Locate an area where rainwater runs off a hard standing - like a driveway, patio or path - and straight to a storm drain.
2. Pick a section immediately adjacent to the hard standing, then look to divert the water flow to this point by either removing edging, bricks or paving; or, by creating rills and gullies.
3. Dig a wide, saucer-shaped pit in the section and use the soil to form a berm around the edge, making sure the pit sits below the hard-standing level, then cut a channel in the berm.
4. Line the bottom half of the pit with gravel, add a gritty soil/compost mix, then plant up with wet-tolerant bog plants, such as yellow flag iris (*Iris pseudacorus*) and meadowsweet (*Filipendula ulmaria*).

The run-off from the hard surface has been deliberately diverted to the rain garden. Photo by BrianAsh/Wikimedia Commons.

exactly where it falls; for example, if you prune a tree the cuttings will fall at the base of the tree. Or, if you grow plants like borage (*Borago officinalis*) or comfrey (*Symphytum officinale*), grow them deliberately under hungry plants like fruit trees. The leaves left on the ground will act as mulch and food for the soil. Whichever method you employ, mulching is an essential art that works alongside rainwater harvesting techniques and is a fundamental requirement in ecological gardening.

Grey-Water Renewal

When exploring the ways and means by which we can harvest the rain that falls in the garden, it's easy to overlook the fact that enormous quantities of freshwater are pumped into our homes, used and drained away every day. Even if we could harvest enough rainwater to sustain the water needs of an entire household, where this water goes after use would still need considering. In the Soil chapter, we covered the possibilities of a compost toilet – which negates the need for dealing with water polluted with human waste – but what about other water used in the house, such as from the bath, shower, washing machine, basins and kitchen sinks? If this grey water was diverted away from the drains, then there is one obvious place for it to go: the garden. In a circular, closed-loop system, there is no such thing as waste – only resources that are currently underutilised and poorly understood. Grey water is exactly that resource. Since it's uncontaminated by the pathogens present in wastewater, it therefore offers us the opportunity to reuse it to grow plants and replenish our watercourses. Implementing the holistic method of grey-water use can be as easy as taking the dish washing-up bowl to the garden, or as complex as replumbing and water manipulation. However, over time all endeavours will be worth the effort, setting a system in place that will be sustainable for generations to come.

The stigma that grey water is unsanitary prevents most people from reusing it, and it can be hard to overcome the feelings of caution and doubt seared into our consciousness. But grey-water recycling is really no different from any other composting method, like decomposing animal manures or fermenting fish bones. Although grey water includes every product, foodstuff, and liquid you have thrown down the drain, it's remarkable how much of this the natural world can absorb and use to its greater benefit. Of course some, like bleach and other poisons, are just too harmful or difficult to break down through grey-water filtration and should be avoided or drained only to the mains system. Likewise, grease, fats and oils used in the kitchen require a grease trap or diverting back to the mains system. However, when the decision to reuse grey water raises your awareness that all the products you use to clean yourself, your clothes, your dishes and your houses will end up back in your garden, it can spur you to be more conscious of the harmful effects they pose and be more inclined to choose only natural alternatives.

Recycling Systems

To set up a grey-water recycling system, the first thing to do is divert all the pipes from the areas you want to recycle water from, so that the water flows out into the garden instead of into the mains drain. This can be the most complex undertaking of the whole operation, and you should hire a professional plumber if you don't feel confident enough to undertake this yourself. Simplistically, this will mean running a single pipe out to the garden that will carry water from the indoor drains whenever they are being used.

At this point, the water will need to be treated to remove the impurities and excessive nutrients. And excessive nutrients are possibly the best way to define what grey water comprises in order for us to use it wisely in the garden. In farming, the corollary would be chemical fertilisers in a non-organic farmer's field. These leach excess nitrogen and phosphorus into the watercourses, which causes eutrophication – when a body of water becomes overly enriched with nutrients, leading to harmful algal blooms. This is exactly what would happen if we sent grey water straight into the garden. So if you want to use grey water, then you need to remove the excess nutrients, and the best way to do that is by partnering with the same class of decomposers as in compost piles: microorganisms.

The difference with grey water is that the material you're trying to break down is suspended in water and needs to be collected in pools for that purpose. Since large pools of still, open grey water are anaerobic, and create odours like any wastewater, you'll need to enlist the help of plants. Aquatic plants will contribute the oxygen needed to turn the process aerobic and speed up the decomposing activity, as more microbes can work in that environment. The nutrients present in the water will also need a medium to attach to and travel through, and you can provide this with either organic material, like bark mulch or compost, or inert mediums such as sand, gravel and rocks – or all these elements in conjunction. Once you've understood the basic process, you can plan how best to hold and move the water through the garden and for what purpose, whether growing plants or returning it back to the watercourses.

There are two main choices: either moving the grey water through a series of large tubs or containers, linked to each other by piping and overflowing from one into the next; or creating a series of ponds or swales that will hold the water and slowly release it back into the ground. The difference between the two is that the containers are watertight and use rock and gravel as the filtering medium, whereas ponds and swales use soil and organic matter as the filtering medium. It's also much easier to collect and use filtered water from the tubs or containers, as they are able to be fitted with taps and hoses, whereas ponds and swales are better suited to large-scale irrigation like flood irrigation or slow-release absorption.

When ponds or swales are used as grey-water filtering bodies, they need rather more drastic landscaping, and we know they work best when built on the contour of a downward incline, allowing gravity to force the water through the system. Your outlet pipe can run straight into the swale or pond, and the entire area will need to be heavily planted with aquatic plants and wet-tolerant trees. Plants such as reed (*Phragmites australis*), spike-rush (*Eleocharis palustris*) and yellow flag iris (*Iris*

MAKING A BATHTUB REED BED

To acquire an old 1901 bathtub, search reclamation yards, house-clearance auctions and online auctions. In addition to a used bathtub, you will need to source at least two barrow-loads of each material: small gravel, larger rocks and stones, and a woody/gritty compost mix.

1. Seal the plughole of the bathtub, then fill up to a third of it with a layer of small gravel (figure 1).
2. Secure three baffles across the width of the tub, reaching down to the top of the gravel layer (two baffles are shown in figure 2). Fill with a layer of larger rocks and stones, two-thirds the way up. The baffles will prevent the water from flowing straight over the top and through the tub, instead forcing it down into the gravel.
3. Fill the final third with a layer of woody/gritty compost mix and

plant the reeds in the compost at even spacing.

4. Fit the grey-water outflow pipe to come into the bathtub from the opposite end to the overflow drain, making sure the compost level is above the overflow drain (figure 3). You can fit a tap to the outside of the overflow drain, or allow the water to percolate out to a soak-away, or fit another pipe and run the water out to a pond, swale or another container.

In this system, the rocks, gravel, soil and roots of the plants act as a filter that suspends the nutrients and toxins, and allows the microbes to break them down. The mulch acts as a cap over the water, trapping odours and hiding the water until it's perfectly clean - a major advantage. Once the water reaches the bottom of the system, it can be stored in larger containers or ponds, or be allowed to leach out into the soil through rain gardens or soak-aways.

3

BUILDING A KEYHOLE GARDEN

Keyhole garden materials list:

- Reclaimed bricks or walling stone
- Rubble rocks
- Old/rusty tin cans
- Recycled gravel
- Garden soil
- Matured compost
- Hazel polestakes
- Thatch - thin reed or straw
- Cordage/twine
- Straw
- Wood ash
- Fresh compost ingredients

The keyhole garden's growing medium is layered, creating a basic filter. When the grey water is poured into the funnel, excess nutrients and toxins will be taken up by microbes present in the fermenting fresh compost and will then be further filtered by the funnel walls and the aggregates at the bottom of the growing medium. The clean water will be taken up by the plants through capillary action, or wicking, from below - by far the most efficient watering method.

1. Mark out two circles on the ground, one inside the other, with radii of 25-45cm and 100-150cm respectively.
2. Lay walling stone or reclaimed bricks along the outer ring, but with a wedge shape, 50-80cm wide, pointing towards the inner circle at one point.
3. Hammer approx. 8-10 hazel pole stakes (1.2-1.4m-high) into the ground along the inner circle. Weave them together with bark cordage or thick twine, then surround them with thatch - thin reed or straw - to make a basket.
4. Fill the bottom of the inner basket with rocks and gravel, then successive layers of soil, straw and matured compost until full. Then continue to build the wall up to 50-80cm high - either in dry-stack style (no mortar) or with a mud mortar.

5. Fill the bottom of the main bed with a layer of rocks, gravel and rusty tin cans, then successive layers of soil, straw and matured compost - with added wood ash - sloping it towards the inner basket.
6. Plant up the main bed and continue to add matured compost layers on top, adding fresh compost ingredients to the inner basket as needed.
7. Pour grey water only into the inner basket, and clear out and clean the inside of the inner basket from time to time.

An example of a low-lying keyhole garden with banana leaves used as the filter. The keyhole design allows easy access to the inner basket. Courtesy of Sheila Halder.

pseudacorus) can be planted directly in the water, while you can plant the berms with crack willow (*Salix fragilis*) or alder (*Alnus glutinosa*). Each swale or pond can then link to a further overflow pipe or be allowed to move naturally through the system via gravity-led design. This means the water will be kept at the surface until the plants have colonised the area, possibly causing odours, but this is a small price to pay for a natural wetland environment that can be left alone to allow nature to follow its own course and determine its own future.

Stand-alone Concepts

If the whole process of setting up a complete grey-water recycling system seems too complicated or elaborate, there are other, more simplistic methods of reusing grey water. If you don't want to divert your drains and pipes, you can use basins to gather water from washing your hands or cleaning your teeth, or to collect the water from a bath or shower. You can then take this water out into the garden by hand and put it into structures such as a keyhole garden or a hessian-sack vertical garden (see Building a Keyhole Garden and Creating a Hessian-Sack Vertical Garden).

Historically, many of the most adaptive and ingenious methods of freshwater collection, use and reuse came from the peoples and areas of the world where rains were infrequent and groundwater scarce. Take, for example, the Native American Zuni people of New Mexico, who developed sunken beds called 'waffle' gardens to capture and store every drop of rain that fell on their arid desert plateau. These 'waffles' are berms (mounds of soil) that are raised up around individual plants to hold water around their roots; otherwise that water would wash away as quickly as it falls. A similar method, known as 'zai pits', is still practised to this day in the dry northern farmlands of Burkina Faso. Even more ancient were the 'waru waru' gardens of the Peruvian Andes, where flash floods and prolonged droughts made the growing of crops extremely challenging. These communities dug large channels around raised beds, allowing the heavy rains to be held back to then irrigate the beds in the dry months.

That these people did, and many still do, raise gardens in challenging and hostile environments with rainwater-harvesting techniques is profoundly inspiring. If we follow their lead and, even with more constant rainfall, adapt our methods to make every use of this resource, we will respect the vitality and preciousness of freshwater. As our climate is now on a course that will change it beyond all recognition, then the future patterns of seasonal weather will not follow the clockwork regularity of the past. By learning from those who lived and still live in what we might consider extreme conditions, we can gain the knowledge and wisdom to guide our gardens through the shifting sands that lie ahead.

And yet our gardens are more than just the resources needed to sustain them; they express our joy and wonder at the beauty of nature and the natural world. By making them more in tune with the natural order and by following its wild rhythms, they will become greater than just reverent mimicry; they will become natural ecosystems in their own right. Harvesting, guiding and reusing freshwater are key tenets to adopt in order to achieve that stated aim. Ecological gardening can only work when a garden goes beyond the superficiality of aesthetic or productive aspirations and rebuilds the natural bonds that have been depleted, making them thrive again. Freshwater is the lifeblood of the garden, migrating and moving through the landscape in perpetuity; and it is the very means of survival for plants and the bringer of all life. When our gardens become sustained only by freshwater we harvested or reused, then we will have made them robust enough to carry on far beyond our own lifetimes; and that is the true test of a good shepherd, a good gardener, tending a natural space.

Produced by
1

2

3

4

CREATING A HESSIAN-SACK VERTICAL GARDEN

Another small, and less permanent solution for reusing grey water, in a similar vein to the keyhole design, is a vertical sack garden. You can make these from used tall hessian coffee sacks or grain sacks with an impermeable drainage pipe placed in the middle.

1. Acquire some used large hessian coffee bags and a piece of 30–40mm conduit pipe, and cut the pipe to the height of the bag. Most recycled coffee sacks can be sourced online or from a wholesaler (figure 1).
2. Fill the bottom quarter of the sack with large rocks, small stones and gravel: large rocks at the bottom, then small stones and finally gravel on top.
3. Secure the conduit pipe in the middle of the bag and fill it with char and gravel.
4. Fill the rest of the bag with soil and compost with added wood ash. Secure 3–4 stakes around the edge and fasten to the bag with twine.
5. Make cuts in the bag down to the gravel layer, then place plants in the top and the side holes. Cutting holes in the side will increase the surface area for planting (figure 2).
6. Pour the grey water into the conduit pipe filled with char and gravel (figure 3).
7. This design allows the plants to draw the purified water up from the bottom through capillary action, or wicking (figure 4).

CHAPTER 5

Wildlife

The wildlife in our gardens is as fundamental and intrinsic to building natural ecosystems as our beloved plants. Although we do not actively release bumblebees or dragonflies into the open air and we may not envisage we are making a home for shrews and slow-worms when we take care of our hedgerows and apply compost, if we undertake the right actions, we will attract more birds, more insects, and more mammals to our gardens. Wildlife perfectly illustrates that no matter how much we think we own our green spaces, we do not. Nature doesn't recognise boundaries or demarcations; it only sees another green space, under the same blue sky. Wildlife is exactly as its name suggests: vital life in the wild form, undomesticated by the human hand.

I vividly remember spending hours as a child lying and watching, absorbed in the endless industry of a colony of ants traversing their way across the patio. Or trying, and mostly failing, to coax woodlice out of their subterranean slumber and into an old shoebox so I could watch them more closely. As I grew to spend my life in the garden, I ended up unintentionally disturbing the lives of many a poor worm, centipede, frog and toad. However, as I gently carried them back to the quieter corners of the garden, I was enriched by the experience of holding them in my hands and studying their forms up close. Many of the more exceptional moments – like standing underneath a swirling murmuration of starlings close enough to touch, or coming across a sleeping bat, all alone, hanging upended on the slender edge of an ivy leaf – are few and far between, but all happened while in the garden and all will be remembered forever.

Once we understand the complex webs of sustenance and survival developed by the myriad of different species that exist in the garden, we can have confidence to implement actions and methods that will only bring about positive effects. Our goal as ecological gardeners is to provide habitat that wildlife species recognise in the wild, not only for their sustenance but also for sanctuary and places to raise their broods or hibernate through the winter months. Much of wildlife is nocturnal, and more goes on in the deep of the night than we will ever know. But we can aid their moonlight industry by allowing their paths and tracks to be unimpeded, and by creating open boundaries and wildlife corridors throughout the garden. Without wildlife, our plants would survive but they would be a dull palette of monotone forms, as wildlife, through pollination, is the driver of the rich tapestry that we see in the colours and forms of plants today. By encouraging and fostering the pollinators' needs, we will avail this symbiosis. There is so much we can do. How joyful to catch a fleeting glimpse of the technicolour coat of a bird or the

To construct its silken web, a spider uses its own body to measure lengths.

acrylic spills of a butterfly's wings; such sights enrich our sense of wonder and affinity with the world. The more we do to encourage and regenerate this mosaic of life, the more we will benefit ourselves as a result.

Web of Life

To better aid and encourage the biodiversity of our garden, we need to understand how food functions at the core of a biodiverse ecosystem. Food drives the waking hours of almost all wildlife species, foraging to sustain themselves, feed their young or collect for a community group. Some travel hundreds of miles to reach feeding grounds, and many risk their lives to guard a food source from interlopers. Every creature, from a butterfly to a hedgehog, has spent millennia trying to carve out its own idiosyncratic source of sustenance, as part of the cycle that we observe as a food web.

Although the food web is a *web* and non-linear, let's begin in the soil. After all, the soil is where you can have the most influence on the biodiversity of your garden. By adding organic matter into the soil on a regular basis in the form of compost, you will do infinitely more to aid the food web than anything else. The decomposition of organic matter is like the starting gun for wildlife creation, and we must play our role at the beginning of the race. Soil supports life by providing sustenance to plants, which in turn provide food and shelter to invertebrates, birds and other creatures. In addition, the microorganisms – bacteria and fungi – that decompose matter and make all this possible are also the foundational food source for another level of consumers: the soil dwellers.

While the results of their activity are indisputable, we cannot see microorganisms with the naked eye. The soil dwellers on the other hand – species including earthworms, millipedes, centipedes, springtails and bristletails – represent the moment where the food web enters our visual realm. Above ground, the herbivores that feed on the plants growing in the soil represent another level. Invertebrates such as aphids, whitefly and the larvae of

The lesser horseshoe bat, clinging onto the edge of an ivy leaf.

butterflies and moths all consume plants from their leaves to their roots, and they in turn become food to another level of consumers, comprising both carnivores and omnivores. I cannot describe the complexity of life beyond this point in the terms of linear ascent, other than to say that these in turn may get eaten by more carnivores before we reach the apex predators at the top, who know no fear of consumption. This is precisely because all these complex species interactions form a web, not a linear chain. The badger is a familiar example of a species that happily dines across multiple trophic levels, from the bulbs of plants and soil-dwelling earthworms to carnivorous moles and frogs. Although it would be bewilderingly complex to try and map out who eats whom in the messy middle of this system, where there is simplicity we must embrace it. So large species of mammals, birds and reptiles, such as a fox, sparrowhawk and grass snake, are likely to be the top predators in your garden ecosystem and are as much the 'end' of the web as microorganisms are the 'beginning'.

Understanding the myriad of relationships that exist within the food web is less about pigeonholing wildlife into convenient boxes than using this knowledge to further our goal in creating garden ecosystems brimming with biodiversity and life. Relationships between species cast a light on what and how we can plant to bring about the most complexity. One useful way of looking at how species interact in the food web is to sort them into either generalist consumers or specialists. Both, however, are equally important and both equally contribute to the balance of the food web system.

Generalists are animals that have relationships with a wide variety of plant and animal species, as they have evolved to eat a range of available nourishment. Many of our resident bird species in the UK, for example, are archetypal generalists. Robins spend much of the year eating a range of insects and other invertebrates, including worms and spiders, but switch to fruit and berries across the barren months, when their choices are more limited. The blue tit also prefers insects and spiders, but will eat seeds outside the breeding months. Many spiders too consume a variety of prey, taking a wide range of flying and ground-dwelling insects, while frogs and toads have a similar lack of compunction about what they eat, devouring an array of slugs, snails, worms and insects.

Specialists, however, are those species that have developed an affinity for only a very limited food source, sometimes even down to one particular species, and without it they would cease to exist, including the brimstone butterfly (alder buckthorn, *Frangula alnus*), convolvulus hawk-moth (field bindweed, *Convolvulus arvensis*) and Robin's pincushion gall (aka rose bedeguar gall) (dog rose, *Rosa canina*). In fact, many of these specialised relationships are so well known that their common names reflect the affinity between host and consumer, illustrating a perennial bond that goes far back through the millennia, from the gooseberry sawfly (gooseberry, *Ribes uva-crispa*) to the hawthorn shieldbug (hawthorn, *Crataegus monogyna*).

For a biodiverse garden ecosystem, it's important to plant species that will attract a balance of both generalist and specialist species. While encouraging the specialist eaters alone may seem to be an overwhelming, complex venture, there are some patterns that can simplify things a little and lead the way. There's a correlation between many of the specialists and some of the most injurious (to some) native plant species. Nettles (*Urtica* spp.), for example, are by far the most favoured plant for the larvae of many butterfly and moth species, while bindweed, bramble and ivy have many special relationships with both larvae and the adults. This widens out to include a spectrum of native plant species that, loved or loathed, have formed the most distinctive and specialised relationships with the native wildlife. It would seem only common sense that the longer a plant species has been present within a geographical region, the longer a wildlife species has had the chance to evolve alongside it. Consequently, the best thing we can do to aid this diversity is to plant as many indigenous plants as possible and especially the less regarded, such as nettles, bindweed and brambles.

The strategy to attract generalist consumers, however, is more about providing enough food through the whole year, like planting a mix of plants, including those that bear berries, nuts and seeds late in the season – such as spindle, hazel and teasel – and those that offer nectar-rich flowers through the spring and summer months, such as crab apple, foxgloves and hemp agrimony. In addition, it can be about providing the base of microorganism-rich soil that will produce the wide variety of life the generalists can feed on – which once again shows the importance of regularly adding compost to the soil. This can also highlight how certain

Aphids. Amazingly, the female's offspring can already have offspring within them at birth. Photo by Andy Nowak/iPhoto.

species become food for the widest variety of life and play important roles not just for the general consumers but for the entire garden ecosystem. I would put worms, aphids and ants in this category, as healthy populations of these species usually correspond to an abundance of life within the whole food web. Worms thrive or not, depending on the amount of organic matter and microorganisms in the soil, which we can control and provide, while aphids can be left well alone instead of bearing the persecution that we normally wreak on them. Ants protect the aphids from over-harvesting and are themselves an important food for multiple species. Again, fostering greater biodiversity can be more about what we don't do, as opposed to what we do, and we must learn to leave certain species alone to allow their populations to flourish.

I probably see worms more regularly in the garden than any other species through the working year. Their abundance in the compost pile and on the surface of the soil means I come into contact with their environment the most. However, I will also seek them out, as their presence is such a good indicator of soil health. If I'm visiting a garden for the first time or checking on its progress after some soil improvement, I will search for worms to guide me in judging the health and prosperity of the soil. Their action in the compost too can be an indicator that a heap is moving away from the thermophilic stage and either needs aeration or is ready for the garden.

Open Boundaries

A garden can be open and inclusive or it can be private and secluded, and yet what it cannot be, if we hope to harbour a healthy wildlife population, is enclosed. The etymology of the word 'garden' refers to 'enclosure', but maybe that's why I think we should strive to create ecosystems instead of gardens. To enclose, to me, is to shut out, to barricade and restrict all things from all sides. If we do this, we shut out a vital element in the wildlife ecosystem: the mammals. Badgers, hedgehogs, shrews, voles, mice and foxes all play important roles within the ecosystem, and if they're restricted, the consequences will be keenly felt. Without these high-level predators, populations within the web of their prey will go unchecked and become unbalanced. Slugs could become more prominent without hedgehogs to feed on them, while frogs may increase exponentially without badgers and foxes to keep them in check. Many, if not all, of these mammals live out their natural habit under the cover of darkness, and much like most wildlife, we'll not see or even notice their habitual excursions. To block them out seems just cruel and mean-spirited, and we can look to arrange our garden to encourage them, not repel them.

Wildlife Corridors

The best way to invite mammals to the garden is to think like one. Smaller mammals, like shrews, voles and mice, travel through tunnels and trails under the cover of vegetation. They need a way in, but a way out too, and a corridor to traverse across the garden without being seen by large predators. Moles use the soil to hide underneath, while squirrels use the trees. Hedgehogs are just as secretive as shrews and voles, and travel through hedges and thick vegetation, while the large dominant species – badgers and foxes – have less fear but shy away from open spaces. What this illustrates is the need for vegetative corridors running throughout the garden without large gaps or open areas in between. So native hedgerows can run internally from the garden boundary to wildflower

The native red squirrel going about its business in the shelter of a naturally formed tunnel. Courtesy of Sam Chislett.

meadows or areas of long grass. Or herbaceous borders can end at a hedgerow and begin by the long grass, and so on. In addition, the hedgerows, or long grass that runs to the boundary, could link to either a neighbour's hedge or out to woodland beyond. This creates even longer corridors, reaching further out into the wild and increasing the chances of mammals travelling along them. The natural habitat that these mammals rely on is forever receding, so we must try to do more than just make our boundaries more permeable; we must also give them the pathways and environments with which to follow their natural patterns and manner.

Birds

If there is one class of animals whose existence emphasises the futility of trying to restrict the free movement of wildlife, we need look no further than birds. Boundaries don't exist to them and they personify the idea that gardens should not be separate parcels of private land but tiles in a mosaic of natural spaces, linked across the country. Birds buck the trend of most other wildlife, as they can be both visually and sonically conspicuous. Their acrobatic swoops and pitches inspire great wonder, while their songs often become the soundtrack of our lives in the garden. Many gardeners, including myself, can attest to having a 'friend' by their side as they work, in the form of the humble robin. Robins evolved to follow behind boars and other large animals, whose snuffling and hoof prints disturbed insects for them to snack on. Gardeners play a similar role for them, our activities lifting up the tasty worms for more easy pickings as we go about our business. It's a fleeting moment of profound joy to look a robin straight in the eye, barely three feet away, before it disappears in a blink to perch back in the branches. Of course, if the robin were the only bird species in our gardens, we would have a poor diversity of life, and there are many things we can do to attract a vast array of bird life into our green spaces.

Birds can be roughly sorted into two categories: those that are resident and those that are migratory. How we encourage each group differs slightly, but many themes remain the same. Birds

Greenfinches are resident birds visiting the garden for seed or habitat in the hedgerow. Courtesy of Sam Chislett.

have a wide and diverse range of habitats, but tall trees, dense overgrowth, and tall grasses and perennials cover the majority of their preferred spaces. If we can be sure to provide some mature trees, left unpruned, with a hedgerow of native species and areas of wildflower meadow, nettles and bramble, we will give the perfect environment for populations to thrive. In the main, this will provide the food source for the birds' vivacious appetites, with vast quantities of insects and other invertebrates needed to sustain themselves and their broods. Summer migratory species such as swifts and swallows will easily catch thousands of insects in a single day, playing a vital role in the balancing of the food web system. Residents like tits, sparrows and finches need these summer populations too, but crucially they need food all year round and rely on the seeds and berries through the winter months. Hence why native hedgerows with berry-producing shrubs and trees like hawthorn, rose and dogwood are an essential habitat, and also why letting perennials, biennials and annuals go to seed is so important. And we must not forget that birds need a drink too, so having ponds or areas of standing water will encourage species even more and provide the full habitat they require. Resident UK bird species that are currently under threat include the song thrush, starling, house sparrow, willow tit and yellowhammer.

Wild Habitat

Birds not only require food, however; they'll need nesting and roosting sites, and we can say the same of almost all wildlife in the garden. Shelter and sanctuary are just as important for the ecological gardener to consider in forming a garden with the potential to house diverse life. Many species will, of course, need completely distinct places to nest in, and most will choose their site far beyond anything we try to provide, but subtle differences in the way we garden can aid their search. This can include increasing the amount of decomposing wood in the garden, through constructing decaying log piles or keeping plenty of ageing mature trees standing with some rotten wood within them. Piles of damp, decomposing wood provide the perfect cover for creatures such as beetles, snails, ants and even voles to nest and breed, while mature trees with rotting branches provide suitable sites for nests of many species of birds, but especially woodpeckers, blue tits and nuthatches. Compost piles too can be a great site for slow-worms, centipedes and toads to either hibernate or reproduce, while ponds and bog areas are the breeding ground for frogs, grass snakes and mayflies.

Although creating natural habitats will always be the best way to encourage wildlife to nest and roost, this isn't always possible, and it can sometimes pay to construct artificial sites to assist certain species. Bat boxes, hedgehog houses, bug palaces and nesting boxes for birds can all be constructed using reclaimed materials, and will increase the likelihood of these species seeking shelter in the garden. It doesn't mean it's guaranteed, however, and it also means that some of these shelters will need maintaining and cleaning annually. But in times of great need, which many species are currently experiencing, a perseverance and dedication to the cause of reversing their decline and degeneration is a worthy aspiration.

BUILDING A PALLET BUG PALACE

Bug palaces are excellent ways to encourage more invertebrates to your garden by providing habitat. This design uses reclaimed wooden pallets, filled with a variety of materials, including:

- Straw
- Twigs
- Hollow plant stems
- Bricks
- Rubble stone
- Logs
- Old timber
- Pine cones
- Bamboo canes
- Cardboard
- Bark
- Soil/compost

From my experience, it can be surprising how much material you'll need to fill the bug palace. Harvest as much as you can from the surrounding countryside and have a sizeable amount of soil and compost in reserve for the backfill. It can be quite a task to complete on your own so is far better if done with others, especially children. There's no better way to demonstrate the habitats of various wildlife species, and how they will benefit from the palace, while doing

something fun and crafty at the same time. Place the palace where it can stay put for many years, as the longer it remains the more bug life will come to populate it.

1. Source five recycled wooden pallets. You can easily source pallets from retail outlets that acquire a surplus (figure 1).
2. Drill holes in the masonry (bricks, stone) and the wood (old timber, logs). Try to drill holes of different sizes, as this is more likely to attract bees (figure 2).
3. Place the first pallet on level ground. Begin filling in the sides of the pallets with an assortment of the sourced materials, then backfill the middle with bark compost. The backfill makes the bug palace secure and creates habitat for soil dwellers (figure 3).
4. Stack another pallet on top of the first one and continue filling in, then repeat until all pallets have been used. Be sure to pack the materials in tightly (figure 4).
5. Cover the top with reclaimed boards and weigh down with stones, then screw a sign to the top. The bug palace should last for many years and see many visitors (figure 5)!

4

5

Pond Life

Of all the habitats we can create, none are as effective at enhancing your garden's biodiversity as a pond. Wildlife species use ponds not only to breed and spawn but also to hunt and take a drink, while some creatures use the mud created by ponds for nests and others will roost and take refuge in its vegetation. A wildlife pond is not a fishpond, however, and it's also not for ornamental aesthetics alone – although it will always be beautiful in its natural state. It can certainly be used as a water source for garden irrigation, and most are integral to water harvesting designs, but this may be impractical for the very smallest ponds. They can be dug into the ground or contained in barrels and troughs, and they can be dug high on a hillside or at the bottom of low depressions. It doesn't matter where they go; as long as they are filled with water and surrounded by vegetation, they will become magnets for wildlife.

Making ponds in barrels and troughs is fairly self-explanatory, so we can leave that aside and look at how to make a natural wildlife pond. The first thing to consider is a water source. Without a fairly continuous water source, a pond will probably dry out and need topping up manually, which rarely makes sense, so a water source is vital. The pond will be part of a wider network of the garden water system so can be fed from several sources. This includes: rainwater collected from roofs; swales and ditches dug in the ground; watercourses such as rivers and streams or underground springs; and filtered grey water. Once this has been considered, then the overflow of the pond becomes just as important, as we must always consider what will happen in flood conditions. We can release the overflow via a pipe set to a certain height (the overflow level), or run it out naturally over a ramp into a bog or rain garden, or into another system of swales. A last consideration will be the siting of the pond, with the most beneficial being where water has already been observed to accumulate in heavy rains or where a natural spring is present. Once all these permutations have been considered, it's time to dig (see Creating a Natural Clay-Lined Pond on page 160).

Natural ponds are thick with vegetation, allowing for the most varied habitat.

CREATING A NATURAL CLAY-LINED POND

Firstly, you'll need to source some clay. I've had luck with building sites, recycling sites, the upturned roots of trees in woodlands, and in the steep banks of small rivers. A barrow-load will be enough for a small pond, such as demonstrated here, with more needed for bigger ponds. Most of the digging can be done by hand, with a spade, but bigger ponds will require heavy machinery like diggers and dumpers. Small ponds can be completed alone, but bigger ponds will need extra help from friends and family to share the workload.

1. Excavate the ground to the desired depth, length and width, adding an extra 50–150mm of depth to allow for the clay, depending on the size of the pond. Ponds are best dug with a shallow beach at one end, allowing for wildlife to get in and out as well as acting as an overspill point in heavy rains (figure 1).
2. Begin puddling the clay (compacting it to remove all air), mixing it with water until it's elastic and smooth. Mix with your feet or with a spade; lots of water will be needed (figure 2).
3. Lay the puddled clay over the bottom of the pond and over the outside berm. If your pond is large, use successive layers of clay.
4. Render-finish the clay with a float, keeping the clay wet at all times, then fill the pond with harvested rainwater and check the overspill height (figure 3).
5. As the pond fills up, plants will begin to colonise and wildlife will soon follow (figure 4).

1

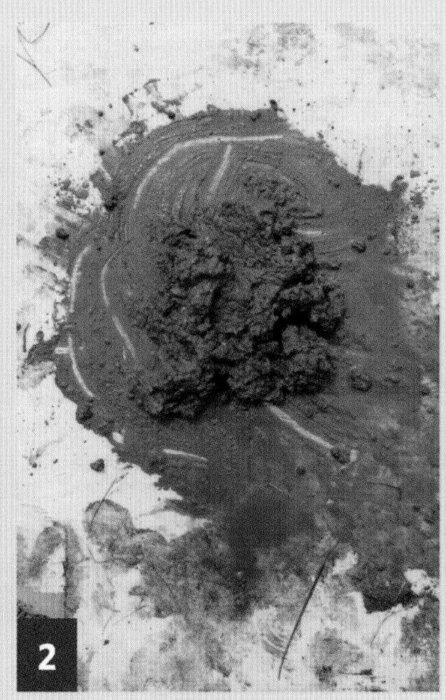

2

3

Once you've constructed your pond and filled it with water, you must now consider the surrounding planting. The newly created berm around the perimeter of the pond – from the dug topsoil – can be seeded with fast-growing annuals like arable wildflowers, or covered in a thick layer of compost and planted up with native perennials at a later date. Don't assume you need water-loving marginals here, such as yellow flag iris or spike-rush, as they won't receive any water from the pond and we don't want any of these plants to get their roots into the pond or use capillary action, which water-loving plants will probably try to do. We are only trying to reinforce the berm to stop it from collapsing in heavy rains. If the pond uses a spillway that feeds into a bog garden, then you can plant this area up immediately. Plants to think of using are meadowsweet, rosebay willowherb and hemp agrimony, all of which play a significant role in habitat creation for the wildlife that depends on the bog wetland environment.

As for planting in the pond itself, I believe it is better to leave this to nature and let it colonise naturally. Wildlife will then undoubtedly follow in kind. If you feel you want to guide the way, then frogbit (*Hydrocharis morsus-ranae*) and water crowfoot (*Ranunculus aquatilis*) will both oxygenate the water and cover the surface to keep the temperature cool and minimise evaporation.

An astonishing variety of wildlife creatures utilise ponds. Insects such as mayflies, damselflies and dragonflies all begin their lives in the water before taking to the skies, while water bugs such as the pond skater and water boatmen spend their entire lives on and under the surface of the water (although some species do fly to other ponds). Amphibians like frogs, toads and newts all breed and raise their broods in the water, while molluscs like water snails play an important ecological role by feeding on the algae that form on the surface. Many mammals too are drawn to the pond for a drink, although which ones and how frequently they visit can depend on access; you can help them by constructing a spillway to a bog or rain garden. Birds too will frequent a pond, whether to drink or to hunt for prey, while swallows and song thrushes in particular will use the wet mud to build their nests.

In brief, ponds can be such important hubs for a diverse web of life that building one could almost be considered a prerequisite for encouraging your garden towards becoming a fledgling ecosystem in its own right. I have always been drawn towards ponds in gardens, much like the wildlife, maybe with the greatest expectation of seeing life – like a school of tadpoles in the water or a dragonfly on a stem. The habits of wildlife are mostly hidden, or fleeting at best, but in a pond we can glimpse their world with sometimes great clarity and detail. It's no coincidence that waterholes are the best place to look for life in the wild, and ponds are no different. We all need water to survive, so if we can provide that in the garden, it will reward us with an abundance of biodiversity.

Pollinating the Planet

Pollination is the thread that binds the two kingdoms of plants and animals together more than any other. Insects and their larvae may eat plant leaves, and mammals and birds use their leaves and twigs for nests, but without pollination almost all the plants we know of today would not be here. While some plants reproduce by vegetative means, most of them reproduce by seed. To create seed, plants need pollination to occur, and many species have found wonderfully inventive ways of doing this, such as pollinating themselves or using wind or water. But the overwhelming majority need insects to carry out this function, and this mutual coexistence between insect and flower has lasted for hundreds of millions of years.

Why then would we, as humble gardeners, have any need or desire to involve ourselves in one of nature's most perfectly harmonious working relationships? Because we, as a human race, already have involved ourselves, through our use of pesticides and destruction of wild habitats, to such a degree that we've weakened and threatened this sacred bond. And yet we occupy a uniquely beneficial position in our green spaces to make a significant difference to this ecological imbalance. We can provide the habitat for the pollinators to thrive and can reap the rewards of the abundant species diversity that they create, in addition to their beauty.

The average lifespan of a butterfly is just one month. Photo by Estuary Pig/Shutterstock.

The pollinators are a varied group, which can include mammals, birds and ground-dwelling insects like beetles and ants, but by far the greatest number of pollinators are flying insects such as bees, butterflies, moths and flies. Within this group it's the bees that are the most ubiquitous in their task, and further still it's the honeybee that grabs the most attention. The honeybee is an effective pollinator, but we must remember that it's the wild bees (bumblebees and solitary bees) that offer the greatest chance of enhancing biodiversity in a garden ecosystem. This is because they themselves are more diverse: here in the UK, we have 24 species of bumblebee, over 240 species of solitary bee, and only one species of honeybee.

This again points to the generalist and specialist relationships that exist in nature. Honeybees are generalists, seeking nectar and pollen from many flowers, while some wild bees are specialists that visit only one plant species. Take, for instance, the harebell carpenter bee. Without a source of native harebells or bellflowers it would struggle to survive, if not die out completely, reducing the diversity of the bee population. Our role, therefore, is to plant the widest range of nectar-producing native plant species possible, to offer a nectar source through the entire year from early spring to late autumn. By delivering not just the food source but also the correct host to that food source, we will provide across the whole range of generalist and specialist pollinators, enabling the diverse ecology to thrive.

Butterflies and Moths

As gardeners the main way we can help pollinators is by supplying sources of food, but we can also provide them with habitats for nesting and raising their offspring. For butterflies and moths, this is more about giving a home to their larvae than anything else, an act that can be of equal, if not greater, importance for their ongoing survival than producing nectar.

Moths do all their pollinating through the night, and since bees are sometimes quite elusive, most people have the greatest affinity for butterflies, their spectacular patterned wings like miniature works of art. The larvae of butterflies are much the same as moths and will benefit from either a nettle bed or areas of brambles and thistles. The adults will be most at home in a species-rich wildflower meadow, however, as the flowering corresponds with the time that they are actively looking for sustenance. So an area dedicated to a native meadow, no matter how small, will always aid the butterfly in its fleeting life cycle.

Wild Bees

Wild bees too benefit from a sustained source of nectar, but unlike moths and butterflies, wild bees – bumblebees and solitary bees – are nest makers. As gardeners we can help provide the habitats needed for their creation, and with solitary bees a whole set of subspecies are defined by these very requirements.

Mason bees prefer to site their nests in small cavities and gaps in walls or stone, using mud to construct the nest and raise their brood. To aid them we can make our walls and steps in the dry-stack method or with only mud mortars that will allow and encourage this species to make their nest there.

MAKING A RAISED NETTLE BED

Moth larvae often have very specific plant-host relationships, depending upon the species, but many, including the mother of pearl and the burnished brass, have a certain fondness towards one specific plant: nettles. Consequently, we can create a dedicated nettle-bed area, to harbour the moth larvae and thereby increase the pollinating moth population.

1. Source some used scaffolding boards and cut two lengths of 90cm and two lengths of 45cm. Then, using 50mm × 50mm stake wood, cut four 20cm lengths. Always measure twice and cut once (figure 1)!
2. In a suitable place, clear the ground and hammer in stakes to half the depth of the boards at each corner, then screw the boards to the stakes.
3. Run chicken wire along the bottom of the bed and nail into place. Use U nails to fix the wire to the board (figure 2).
4. Fill with a layer of soil and a layer of compost, then plant the nettles at even spaces throughout. Cut back to 10–15cm and water thoroughly. It helps to wear gloves when planting the nettles (figure 3).
5. Cut the nettles back in early spring to encourage fresh growth (figure 4).

4

Bumblebees live in colonies of up to 200 workers, with new queens produced every year. Photo by Michael Meijer/iStock.

As their name suggests, **mining bees** 'mine' or burrow their nests into the ground, making elaborate networks of tunnels in loose, sometimes sandy soil. We can assist this preference by leaving areas of sandy soil unmulched and free from vegetation if we notice a nest present.

Carpenter bees like to make a nest in the trunks and branches of trees, or even treated wood poles, benches and planks. A garden full of mature trees and reclaimed timbers will provide the habitat they need, or we can help them along by drilling holes into trees or timbers or by leaving a collection of stacked bamboo on the ground.

Finally, there are the **leaf-cutter bees**, which will take plant leaves to make nests in sites such as the hollow stems of dead plants or similar cavities in wood, masonry or the ground, much like all their related solitary species.

Bumblebees, on the other hand, are not solitary and will make nests either above ground with dead grasses and moss or below ground in old mammal holes. Keeping areas wild and untidy with moss and dead grass present will be the best we can do to assist bumblebees in their nest building.

It's quite astonishing to consider the difference it would make if the pollinators were not quietly going about their business in the garden. There would be no apple, cherry, plum and pear fruit hanging from the trees, while elderberries, blackberries and raspberries would not form on the shrubs and bushes. Rosehips, rowanberries and sloes too would no longer be a source of food for birds, and great swathes of annual and biennial wildflowers would no longer display their colours across the ground. In short, if we like plants we must love pollinators and do everything we can to provide sanctuary and sustenance to aid their health and industry. Ecosystems are always more nuanced and complex than we can hope to imagine, but small differences in how we garden will be hugely beneficial to these populations, allowing nature to recover its vigour and resilience. As long as the pollinators are free to play out their natural life cycles, the biodiversity of the garden ecosystem will be at its most abundant.

'Pests'

It's a given that making and sustaining gardens in a regenerative manner will mean following the principles of organic stewardship. But it needs further emphasis and to be readily embraced when it comes to the wildlife populations we hope to foster. Nature does not understand synthetic chemical pesticides, nor does it want anything to do with them. Throughout my working life I haven't met many wholehearted advocates of chemical control, but I have met many more dabblers in the dark arts. People will say: 'I only put a few slug pellets down' or, 'I just sprayed the black spot on the roses'. But just as small positive actions add up to something much greater when combined, so do individual negative actions accumulate for the worse. Condemning pesticides to the annals of history can only have positive, regenerative outcomes.

Some think of foxes as pests, but they're an indigenous species and should be left alone.

I have always disliked the negative connotation of the label 'pest', much as I have that of the label 'weed'. For the pest is wildlife, and the weed is a wildflower, and the more we step away from this anti-nomenclature the less likely we are to embroil ourselves in negative action. Let's take, for example, the most common garden foe: the slug. It's hard to ask many gardeners to love the slug, but it can be easier to say you don't need to live alongside an overpopulation. As we have seen, slugs are an important food source for slow-worms, newts, moles and ground beetles, among others. So by putting our action into increasing populations of these predators, such as through ponds or decomposing log piles, we will naturally decrease the slug population.

How about the other supposed arch-enemy of the green-fingered: aphids. Aphids, as I mentioned earlier, are an extremely influential species in garden ecology, such is their importance as a food source for so many species in the dynamic food web. And they are, in their adult state, also wonderful to observe, though they are the bane of the veg and rose gardener. So if their numbers are multiplying exponentially, then the only solution is to encourage more ladybirds by providing more adequate hibernating habitat over winter, like log piles and dead plant stems; or by accommodating more swifts by allowing them to nest in the eaves of the house, and more hoverflies by providing them with nectar-rich plants. The idea is never to bring about the end of slug and aphid populations, as this would be futile and detrimental to the overall balance of the ecology. We're only making subtle adjustments to the web and only if we really feel the need, and perhaps it would best if this were never at all.

This does not mean we are not immune from setbacks in our endeavours in the ecological garden – such as fungal infestations, or the over-predation on young seedlings or cuttings by slugs and caterpillars. But in all my years as a gardener, I haven't encountered any problem that wasn't created by shortcomings in method and knowledge, or that couldn't be counteracted by simple means. Fungi, for example, are vitally important to the decomposing and nutrient recycling of dead plants in a balanced ecosystem, but they are also opportunists and can turn to pathogens and parasites of healthy plants too. Black spot, for instance, is a common fungal disease of roses. It can sometimes be the result of poor pruning, whereby a plant is pruned hard and then not pruned again, which causes an overabundance of stems and an unnatural thickening of the bush, making the plant more susceptible to fungal attack due to poor air circulation. Therefore, correct pruning to allow an open bush that lets air through and avoids unnaturally damp conditions in the bush will increase the plant's natural defence. Black spot can also be caused by poor soil health and balance, which can be countered by periodically adding organic matter in the form of compost to the soil. However, in my experience the greater determining factor for black spot is the changing patterns of seasonal weather, as it thrives in damp, wet and humid conditions. So if a year sees a wealth of this weather, then black spot is most likely to form. And if it does, all we need to do is leave the plant alone and accept that it has happened this year but may not happen the next, or prune the plant and compost the cuttings.

'Pests' are fundamental to ecosystems, and although they may not always benefit us as gardeners, we mustn't use non-organic methods to constrict their natural tendencies. Instead, what we can do to mitigate outbreaks and infestations organically is avoid searching for fast growth through excessive soil nutrients in young plants, as this will only weaken their defences in the long run. And we can allow plants to grow strong and slow, with as much exposure to their natural environment – in the soil and climatically – as possible and as early as possible. The same philosophy of restraint applies to watering: instead of excessively watering a plant in the hope it will grow lush and more luxuriant, water only when it truly needs it – and well and deep so it reaches the far tips of its roots. (Some young seedlings and bulbs, when still in pots, will enjoy some form of barrier protection, like a cold frame or cage, and I only like to put them out in the garden when they are large enough to have outgrown a 9cm pot.) With an ecological mindset, we can happily enjoy the pleasures that gardening brings while coexisting in holistic balance with the wildlife all around us.

Coppicing in the winter will keep you fit and warm!

CHAPTER 6

Materials

Every garden is different, not only because each exists in its own unique environment and topography, but also because we too possess complexity and variety, and it's by our expressions, actions and guidance that these pockets of nature are stewarded into what we then regard as a 'garden'. Since the earliest realisations of this bond between human and nature in the cradle of civilisation many thousands of years ago, we have used not just plants but also materials to express this affinity. Elemental, natural materials such as stone and wood would have been used along with clay or mud to form areas to walk and sit in. And with the advent of successive industrial revolutions came processes that developed human-made new materials such as bricks, metals and glass. With few exceptions, our gardens are still dominated by this small band of canonised materials, and understanding how to choose, use and reuse them is central to creating gardens as holistic ecosystems.

Materials used to compose and construct within the garden landscape can reflect both the natural world and the influence of humans within it, much like homes and buildings do.

Consequently, their function within an ecological garden is far from negligible, and we need to be more conscious of the impact their use will have on the environment, within and beyond the garden. We can ask: How do these materials affect the soil, water and wildlife already present in the garden? What is the providence of these materials and how will their procurement impact on the environment? If we change the way we look at the construction of a garden and the materials needed to perform these tasks, then the wider effects can be overwhelmingly positive. Reducing our dependence on mass-produced cheap wood means less demand for vast mono-species timber forests, which could be allowed to rewild to their original state. And by rejecting human-made products like concrete and plastic, huge CO_2-emitting cement works and plastic factories could be drawn down and curtailed. If we also avoid any new natural stone, then gigantic open-pit quarries could be left alone to heal and regenerate from the scars of endless plunder. All of this will lead us towards a future of balance and harmony in the natural world, within the garden and far beyond, and we can achieve it through the simple act of choice and conscious application.

It can sometimes seem easier to understand which materials to avoid – plastic and concrete, for example – than to know which materials we could and should be using. However, we only need to follow the same philosophy that applies throughout the wider context of consumer life – if you can reuse it, recycle it or reclaim it, then do so above everything else. And when obtaining anything new, make sure it comes from naturally sustainable sources. In addition, I believe we must consider the biodegradable lifetime of any material brought into the garden and accept the responsibility for that duration – gardens are natural spaces, after all, and will go back to their natural state in time.

Materials exist in a time frame of hundreds, thousands, even millions of years, and awareness of this engenders a more regenerative mindset, approach and action. If we choose only reclaimed or recycled stone, brick, timber and metal, for example, to make our paths, terraces, walls and arches, then we will be using materials that will have no more impact on the environment or their place of origin. And if we also source or produce our own sustainable materials – like wood harvested from a native coppice or clay mortars from the soil – we will balance our needs and the ability to sustain those needs. We can also abandon all unnecessary use of non-recyclable materials like plastic in our gardening method, creating a closed system without an over-reliance on outside influences. And if we always consider the wildlife and ecological elements already present in the garden – like watercourses, symbiotic plant-to-insect relationships and complex soil processes – before we place or use any materials, we will allow our gardens to work intrinsically within nature and not readily against it.

Reclaim and Recycle

In this world of hyper-consumption and over-production, picking through the debris of this runaway train has been practised and observed for many decades now, and although its benefits are known and obvious, it's still important to pursue and advocate. Natural materials like stone and wood will last hundreds if not thousands of years. In reclaiming them, we're not only seeking to slow the destructive loop of extraction and plunder, but we're also saying there is great beauty and natural elegance in using materials that are weathered and worn by the elements. The crumbling, decadent style of Renaissance gardens like the Villa d'Este in Italy are testament to this, and although they were originally built at great environmental and human cost, they're unrivalled in their beauty as they succumb slowly back to nature. To ape this style, some materials, such as stone cathedral flagstones or ancient hardwood timbers, are highly prized, with a price tag to match, but many are not and command only a fraction of the price. Searching and finding the right materials at the right price is part of the adventure and a point of difference from the easy convenience of buying new.

Some manufactured objects, made from only natural elements, such as bricks, tiles and pots – all made from clay – represent a wealth of reclaimed and recyclable materials that have a multitude of uses in the garden. Kiln-fired clay materials are one of the better achievements in industrialisation, as they're made only from what they will eventually return to – mud. It's entirely possible to make your own clay-based garden bricks, tiles and pots, but they're durable and adaptable materials that will last many hundreds of years, so finding and obtaining reclaimed clay bricks, tiles and pots should be at the top of the agenda. Metals and glass too are also useful materials when reclaimed. Metal usually comes in the form of corrugated iron roofing, iron agricultural troughs, or bespoke iron gates and arches. Glass, usually in the form of windows or sheets, although fragile, will always be highly prized in gardens for greenhouses and cold frames. And it mustn't be forgotten that aggregates like sand and stone gravels, used extensively to create hard-standing areas or for the base in other construction elements, can and should be gained in a recycled form.

Stone

How to use these abundant materials and in what context in the garden can vary enormously, depending on the situation and the desired outcome of their use, but in the main they will cover almost any eventuality. Reclaimed and recycled stone, for example, has a multitude of uses in its novel forms of flagstones, rubble stones, cobbles and gravels.

The devil is in the detail here, as we must consider not just the salvaging of the materials, but also how to apply these elements to the garden. The easiest method with stone is to use it 'dry', as in

Table 6.1. Stone Forms and Uses

Form	Uses
Flagstone	Terraces, paths, seats and steps
Rubble stones	Freestanding walls, retaining walls, raised beds, steps and firepits
Cobbles and pebbles	Edging, paths and courtyard terraces
Gravels	Paths, driveways and terraces

Rubble stone can be acquired from reclamation yards, recycling sites and building sites.

'drystone walling', where freestanding walls are constructed without mortar or where flagstones are placed directly onto the earth. This means they can all be removed with no waste and without using inorganic compounds like cement or concrete. It also means they're permeable, so plant and wildlife can colonise them and move freely in and around them. However, this isn't always possible – for example, if you need a wall to retain a large weight of soil or if a hard-standing area is on soft ground that is liable to sink and give way. If this is the case, then the footings for the walls and terraces can be dug out to the desired depths, the earth rammed, and the area backfilled with natural rubble stone to the height required. Then clay or cob mortars (more of which we will cover later) can be applied to the walls or used under the flagstones to add strength and rigidity to the structures, all without the need for any concrete or cement mortar.

The roof lintels from old barns give pergolas and arbours worn-in character.

Timber

If stone is the most copiously used garden material – and therefore most easily reclaimed – then timber isn't far behind. Reclaimed timbers, like hardwood beams and lintels from houses or barns or decommissioned telegraph poles, are readily available and can be used for uprights and supports in tall structures such as pergolas, greenhouses or potting sheds. They also work as an alternative to stone in retaining walls, raised beds and steps, or in contemporary designs for terraces, where they are placed in the ground and mixed with other materials like gravels and brick. Then there's the planked or planed wood – as opposed to the square or rounded bulky timbers – such as old scaffolding planks, pallets, shiplap cladding and interior flooring, all of which can be recycled into decking, tables, chairs, compost bays, potting shed walls, and a whole multitude of other uses. Also, old hardwood whisky and wine barrels eventually reach the end of their use to liquor makers, and these can be repurposed into superb rainwater-harvesting collectors (as we covered in the Water chapter) or sawn in half and used as rustic containers for plants and trees. Some of these wooden materials, such as centuries-old house timbers, will be highly desirable and will require a considerable outlay; but others, like pallets and telegraph poles, will be less costly and can usually be gained at no cost due to their perception as a nuisance waste.

Metal and Glass

Metal too is an abundant and adaptable material that can be repurposed in the garden: for example, the omnipresent corrugated iron sheets found on agricultural buildings can be reused as roofing for potting sheds and verandas or, on a smaller scale, for tops on compost bays and wood stores. Agriculture also provides us with steel water troughs, which can be reused to store harvested rainwater, or serve as soil planters, nursery holding beds or, most appealingly, as micro ponds to encourage wildlife. Wrought-iron gates and arches can be easily sourced as well, and although they don't adapt well beyond their original purpose, that isn't always a problem if that is what you intend them for, as they will add a sense of elegance and age to any garden. Glass – in frames of wood or metal – can be used to construct greenhouses, potting sheds or the tops of cold frames for hardening off seedlings and cuttings. Old windows are much more robust than you would imagine and far superior to the thin horticultural glass put in many new greenhouses. You may also come across porcelain sinks and bathtubs – made of clay – which can be recycled into micro ponds, soil planters or rainwater harvesters.

One of my favourite tasks is to walk through my local reclamation yards, seeking inspiration and ideas for projects and designs. Their sprawling size shows the vastness of materials seen by others in need of replacement, when I see only treasure and possibility. Their journeys to the yard can be as interesting as their weathered shapes; from great oak timbers saved from the bonfire of rural farmland barns in Romania to the hefty Tudor bricks from dilapidated houses of the gentry, all have a story and a history etched into them.

Reclaimed wrought-iron gates weathered by the elements.

If all these materials retain a certain sense of style and charm, adding to the overall aesthetic quality of the garden, then some materials do the absolute opposite – namely non-recyclable problematic materials like plastic and rubber. In landfills, these cause great environmental damage and therefore it's worthwhile reclaiming and reusing them. Rubber tyres, for example, make a great retaining wall for a pond or filled with soil as planters, while plastic-bottle bricks – bottles stuffed with small, thin pieces of plastic – are useful in any number of situations requiring walls. Plastic bottle tops too are a fun way to make a chair top or small path, and plastic bottles cut in half can be used to grow seedlings. Panels from broken fridges can be used as seed-tray heat covers, and rusty spanners or wrenches make unique shed and gate handles. Repurposing items in your own home can also divert additional waste from ending up in the landfill. Instead of purchasing a new dibber, it's easy to use an old wooden cooking spoon or an old spade or fork handle instead; and before you throw out the inner cardboard of your next toilet roll, consider saving it to use as a seedling pot.

In the main, reusing and recycling household or problematic materials is a crafty and fun exercise that relies on imagination and no small amount of long-term collection and planning. It must never be forgotten, however, that the garden is its own closed system and if non-biodegradable materials are brought in, they must be used until their demise – and with plastic, this can be over 500 years. That's a long-term responsibility, and although you will take them out of damaging landfills for now, can you be sure the next occupant of your garden will look after them as well as you?

Coppicing for Wood

In our lifetimes as gardeners-in-chief of our green spaces, there will undoubtedly be a need for a continuous supply of wood, whether it's to replace badly decayed timbers or posts, or to make more trellis, screens, tools and many other useful objects. The only sustainable way to do this is to purchase wood from an ecologically watertight commercial coppice or, better still, grow your own coppice.

An old ash coppice stool, left to grow out into maturity.

A coppice is a collection of trees – always native – that are planted for the purpose of harvesting their wood over a growing cycle dependent on the species. The tree is cut or pruned right down to its base when still young (different for different species, but generally under five to ten years) in order to harvest the wood for material. The tree will be changed but unaffected and will regrow from the stump with considerable vigour, to be harvested from once again many years later. This technique has been used for thousands of years as a source of timber and was a much more common woodland resource method until the advent of the mono-planted timber forest. Coppices that are integrated within a native forest and guided with thought and care mimic the natural environment created by the cyclical falling of mature trees. Once a tree comes down or a coppice is harvested, the area becomes flooded with light and the forest floor explodes into a regeneration race as species begin the tussle for dominance that leads back to eventual maturity, and the cycle is renewed. Curiously, periodic pruning can elongate the lifespan of some trees, possibly due to the relatively young age of the stems as opposed to the roots, and therefore a coppice can eventually be left alone to grow old after a period of harvesting.

Coppicing is a long-term endeavour, so choose a site with care and consideration for the future and, if space allows, plant a mixture of native tree species that you will be able to harvest at different times. The tree species are of vital importance, as each species responds differently to coppicing and will grow at varying rates of vigour and girth. Willows (*Salix* spp.) and hazel (*Corylus avellana*), for example, are fast growers that will produce soft, malleable thin stems, whereas ash (*Fraxinus excelsior*) and lime (*Tilia* spp.) have slower growth and are better suited to cutting when the stems are thicker and more mature. Choosing a collection of both will be the most beneficial over the long run, but if you were to plant only one species, I would recommend hazel as it produces good material at a fast rate and adapts to any soil situation. However, if you intend to plant willows, alder (*Alnus glutinosa*) or silver birch (*Betula pendula*), then ideally the ground should be more wet than dry, and receive more light than shade. In fact, all your coppice species – ash and also black poplar (*Populus nigra*), another common coppice species – will want plenty of light, especially after cutting, but you must still plant the trees much closer together than you normally would (2m apart). Remember, pioneer species like silver birch, hazel and alder will grow much faster than ash or limes, so when planning the spacings, plant the slow-growers more sparingly and slightly further apart.

When to begin the actual coppicing, or cutting, of the wood will depend on the age you are starting from, but let's assume you're planting the coppice with one-year-old whips, or saplings, in a small area. First, you won't want to cut them for at least four to five years, but once they have reached five to six years of age, you can cut them on a rotation – a few trees one year, then a few more the next. You'll want plenty of light to hit the stumps (or stools) so cut these trees in a group to aid stem regrowth. This will also flood light onto the soil around the stump, enabling more plants to grow and more places for wildlife to inhabit.

Coppicing should only be undertaken in winter when all the nuts and berries have gone, the sap is low so the tree won't 'bleed' and there are no nesting birds in its branches. Cutting in the colder months will encourage the tree to put more vigour into the regrowth the following spring. How you guide your coppice cutting is entirely of your own volition, dependent on your material needs. However, always be mindful of the impact this may have on the wider ecology you have now created, and for that reason, I would advocate small harvests, staggered over the years to engender a more holistic, balanced and ecological approach.

A coppice is, in many ways, pure ecological gardening, as the materials we harvest will regenerate and regrow entirely naturally. If we're looking to nature to provide for us, then we must be more than just sustainable; we must give back as much as we take away. Through guiding the coppice with knowledge and skill, you can increase the biodiversity within your garden and lengthen the lives of the trees you cut. The materials you harvest are still the end goal, however, and the different sizes and shapes of the harvested wood will provide for a multitude of different uses in the garden. Young whippy stems of willow and hazel can be

COPPICING A HAZEL

You don't need to plant multiple trees to start coppicing or reap its benefits. If you lack the space, you can coppice a single tree (I recommend a hazel), which will still produce an abundance of good useful poles and firewood.

1. Choose a hazel with stems at least 2m tall and start cutting from the outside in. Use secateurs, loppers or a saw, depending on the thickness. Make sure the blade is sharp in order to cut the wood cleanly (figure 1).
2. Continue cutting into the centre, making a sloping cut to the outside, leaving approximately 150mm of stem above the core stump. Figure 2 shows the 'stools' left after coppicing.
3. Sort the stems into groups of similar size, bundle with twine and store somewhere dry. Stack the limbs by their different lengths (figure 3).

1

2

3

woven together to form screens or permeable shade roofs for pergolas, as well as trellises or supports for climbing plants to grow up and over. And they can also make the head of the 'witches' broom', or besom, the most evocative of all garden tools. The larger stems or logs can be used as posts and poles to support screens and trellises, or you can make them into table legs and handles for tools. In short, a coppice will be your own source of sustainable timber and what you can do with it will be the result of either your imagination or the garden's needs (or both).

I have known hazel and willow stems to be used as climbing wigwams and teepee trellises for as long as I've been a gardener, but many store-bought structures are instead made of bamboo cane. Although bamboo is a natural material, it's as mass-produced as pine from mono-species forests, and curtailing its use and only relying on materials sourced from within the garden is by far the greater option. It can take quite an inventory of stems and poles to support the climbing plants in the garden once the season gets going, so making sure the coppice will provide enough for your needs at least every two to three years is the key to this.

Tools

I love to wander in the garden, watching the butterflies loop and pitch, hearing the rustle of the leaves and the gentle birdsong, but if I'm there with a task to do, I will no doubt have a tool in my hand. And if I'm going to disturb this tranquil idyll, I will not do it with powerful polluting machinery that unhinges the air with its incessant din. I will use sharp simple tools with forged blades and wooden handles. Because if one thing divides us so entirely from the natural world it's the machine, and gardening has become so in thrall to the machine it seems to have forgotten why it even needs it in the first place.

Garden Machines

Of all the machines used in the garden, the most pervasive would still be the lawnmower. It's a simple fact that if you have a lawn, you're likely to own a lawnmower, and it's even more likely to be petrol-powered. What is also true is that your lawn can be long grass, not short, or a wildflower meadow, or even a chamomile or thyme lawn, or maybe not even a lawn at all. These choices negate the need for a mower of any kind and would immediately increase the biodiversity of the garden and reduce its need for excess watering. Hedge trimmers too have a certain prevalence through the summer months, but this is only down to two distinct garden styles: the single-species hedge and topiary. Both styles offer far less to wildlife than in their unrestricted forms, and mixed hedges allowed to grow and then periodically coppiced require no such machinery. Topiary plants can be cut with shears and it can be pleasurable to do so, but they can become so tightly packed as to be as natural as a plastic bauble. Strimmers and brush cutters are just lazy gardening toys, used to clear and tidy areas that don't need it, and leaf blowers are the invention of a sociopath who decided it was time we sort out the autumn leaves by blasting high-pitched air about the place.

Without these machines the garden will not go unattended. Its growth still needs checking, just at different stages and in a different way. There's not a single job in the garden that a hand tool cannot do, and if the gardens we design are built around this idea, then there's no need to spend any unnecessary effort or energy in their upkeep.

Hand Tools

Most of the tools that we've augmented with extra mechanical power are used for cutting, but before petrol we had the blade and its effectiveness has not diminished. An axe, for example, will always be a wonderfully useful tool and the triumvirate of a hand axe, felling axe and splitting axe better still. They're obviously the most useful for working with trees, whether that be coppicing, felling or pruning, and I would also add the billhook to this category for its ability to clear through tough scrub and for laying hedges.

Scythes and sickles are best used on anything with a softer stem than you'd use a billhook for, like wildflowers or grass. These tools seem to have fallen out of our gardening consciousness, but we are only poorer for it. Sickles are one of the earliest known gardening tools and are superb for clearing grass or wildflowers under trees or in other tight corners. You can get the traditional crescent moon blade or the Japanese version with a shorter curved blade. Scythes, as mentioned in the Plants chapter, are really only for cutting large areas of grass or meadows and take a certain amount of skill and practice to master. However, if you're looking to grow meadows, not lawns, then they are without a doubt the best tool for the job.

On a smaller scale, secateurs, loppers, shears and pocketknives will cut and prune with more precision and detail; secateurs especially are a constant companion of mine. Buy the best quality you can afford and keep them all clean, oiled and razor-sharp. Saws will also be an invaluable tool – particularly a folding saw for tree pruning and a bow saw for the ability to replace the blades.

The tools mentioned here showcase the redundancy of their petrol-powered cousins, but some tools are just never going to be replaced by machines – including the spade, fork, hay fork, hoe, rake, broom, trowel, hay rake, shovel and many others – and all are just as essential. In almost all cases, these tools can be made from wooden handles (ash mainly here in the UK) with hand-forged blades. Aim to buy the best and simplest you can and you won't go far wrong. The most important thing to remember is that a properly maintained tool should last decades, if not an entire lifetime and beyond.

Tool Upkeep

One of the most essential and positive ways to prolong the life of your tools, and subsequently make them easier and more effective to use, is to learn how to sharpen their blades and make or acquire replacement wooden handles. For sharpening, you'll need a natural sharpening stone, either two or three separate blocks or one block with two different levels of coarseness on the top and the bottom. One will be for roughing or grinding the edge and the other for honing and sharpening. The best of these are Japanese waterstones. These stones will need to be worked with water

SHARPENING A BLADE WITH A WATERSTONE

Using a waterstone takes skill and patience as well as concentration - remember, you will make a very sharp blade by the end, so be careful and mindful through the whole process. I find it a most rewarding and relaxing task.

1. Submerge the waterstone in water for 10 minutes. You will know that it's full of water when bubbles stop rising (figure 1).
2. Use a scouring block to clean rust and dirt off the blade. Scour the whole blade until no more rust or dirt can be seen (figure 2).
3. Keeping the waterstone wet, sharpen the blade using the rough grit side first, then the fine. Press down on the bevel of the blade as you work it along the stone (figure 3).
4. Clean the blade with a plant-based oil, working the oil into the blade with an old cloth (figure 4).

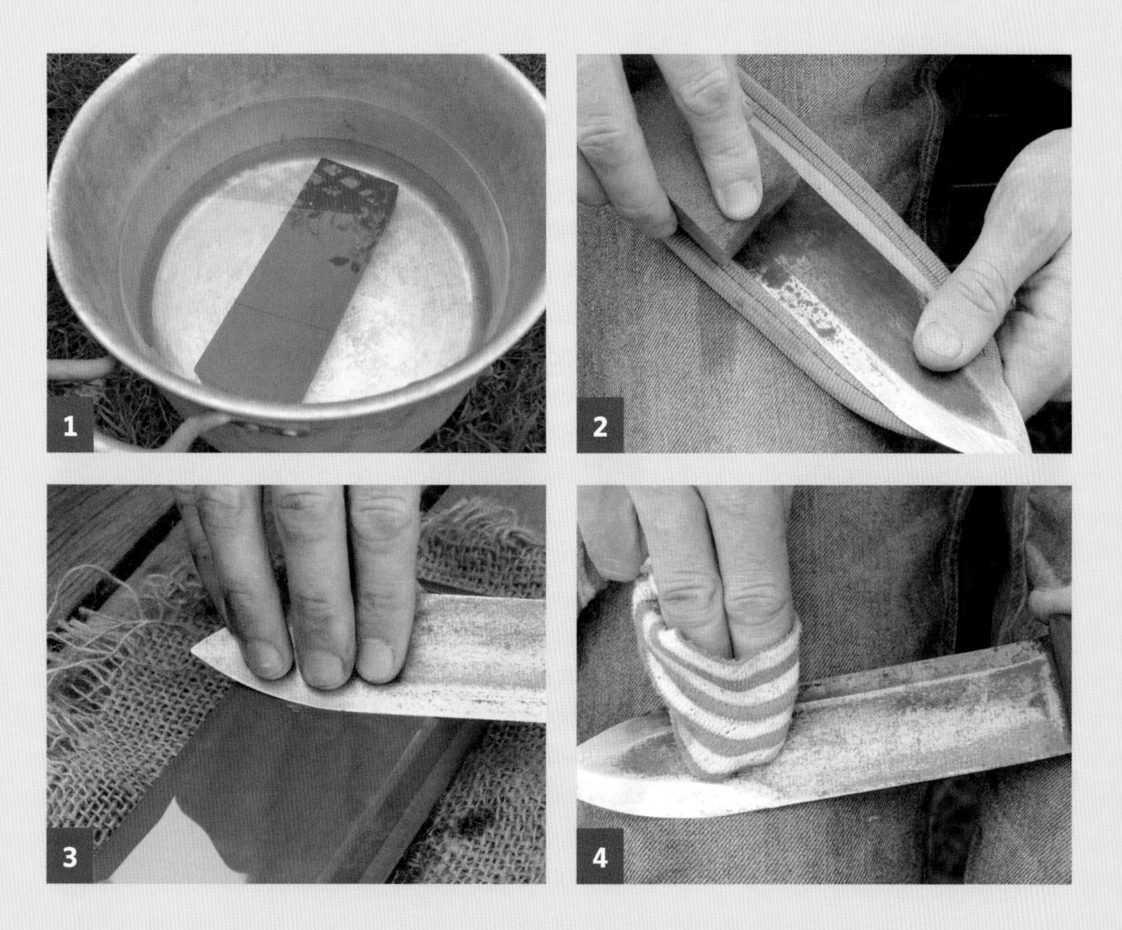

– soaked in water before use and water added during sharpening – and will come as flat blocks that are used on a bench. Or, it's even possible to source your own collection of stones and work them with an abrasive to flatten them down – a process called lapping.

In addition to a 'bench' stone like a Japanese waterstone, you will also need a dry 'pocket' or 'field' stone. These are specially designed for use with the scythe or sickle, as you must sharpen these tools when working with them in the field. The stones usually have a cigar shape that enables you to quickly sharpen the blade, front and back, in a fluid motion up and down the blade. They also work for tools too large for a bench or without detachable parts like a spade or a hoe. Again, these stones should only be natural and it's entirely possible to source your own stone – pebbles are a possibility, as you already have a workable shape. As for the wooden handles, you will be able to source these from your coppice eventually, but until that point make sure they are made from purely sustainable, locally coppiced wood.

I take great pleasure in cleaning, honing and sharpening the blades of my gardening tools. To me, keeping these tools sharp makes a vast difference in doing a task well and with the least amount of hardship. After all, it's the fear of toiling through a job that puts many off the most arduous of gardening tasks, and as a result they reach for the supposed easier option of the powered tool. But I can tell you without reservation that most of the tools in peoples' sheds are nowhere near sharp enough, and if they were, they would be shocked to find how much pleasure they can gain from ditching a weighty apparatus for only a blade in the hand.

No Waste

One of the most distressing aspects of gardening in the modern age is the almost inescapable reach of plastic as the material of choice. From the ubiquitous plastic pot to plastic compost bags and plastic polytunnels, it's ballooned into a monster that is completely at odds with the idea of a garden as a natural space. Somewhere along the line we became indifferent to the harmfulness of materials that were being used to provide our 'green-fingered' pursuits. The plastic pot is as dangerous to the environment as the plastic bag, and eradicating them from circulation and consigning them to the annals of history is the only path to redemption. Within the garden, we can control this by adopting the policy of a closed system to all materials. This means that if plastic comes into the garden in any form, it is our responsibility to use it for the course of its lifetime, and for most plastic this is 500 years. This gives the lie to the idea that we will use plastic pots and bags again, therefore negating their 'single-use' status, as what will happen to them when they break or split and we throw them away? They will, in all likelihood, go to landfill for 495 years instead of 500. All we can do is be more consciously aware of the implications of the decisions we will make.

Some of the most practical actions we can take towards a zero-plastic future can seem obvious when reassessed, like when and how we buy trees or shrubs. If we purchase them from a grower's nursery in the winter, when they go dormant, they will be dug directly from the ground for us to take away as bare-root plants or possibly wrapped in hessian. This isn't the case if we buy them from a retailer, or in the summer months, as the plants will be in pots, not the ground. The same is true for most other plants at any time of year. But with some hessian fabric of our own, we could take plants out of their pots with the soil still attached to their roots and wrap them in the hessian to transport home even in the hottest days of

There are approximately 500 million plastic pots in circulation in the UK alone.

USING HESSIAN WRAPS FOR PLANTS

Hessian, or jute, is an invaluable material in the garden and can be recycled from its commercial use as coffee sacks. With a supply always to hand, its uses in the garden will become apparent - for the plant wraps, just cut some off from the main sack or sheet.

1. Take some hessian and twine to your local garden nursery.
2. Purchase a plant and take it out of the pot. Leave as much soil as naturally accumulates around the roots but leave the rest in the pot. Purchased plants have established root systems, which hold the soil (figure 1).
3. Wrap the roots and soil in the hessian and tie tightly together with twine. Tie the wrap firmly but not too tight as to damage the roots (figure 2).
4. You can now take the plant home just as easily as if it were in a pot (figure 3).

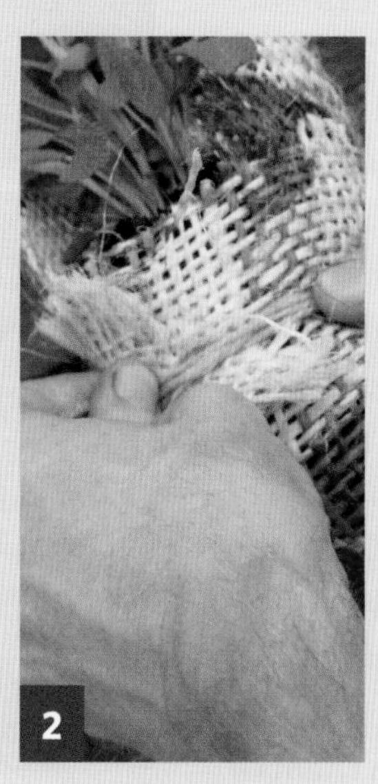

summer. Plants are more concerned about receiving enough water than anything else, so as long as they are watered on the day they're purchased, they will survive their journey to the garden. Buying plants in the middle of summer is not advisable anyway, so buying during the shoulder seasons or the dormant winter makes accumulating unnecessary plastic pots an even simpler task. If we aspire to do this, no more plastic pots will make their way into our gardens, and it might even make the horticultural industry take notice and change their approach to the menace of the plastic pot.

Raising seeds is another common gardening task that needs a fresh approach and new materials. For broadcast seeding, you can easily make trays out of recycled pallet wood or willow or hazel stems woven together, with hessian fabric laid on the base to hold in the soil. For single seeding, paper wraps or old toilet rolls should be readily embraced for their ability to decompose into the soil once their utility has ended. And for raising seedlings or cuttings, there is nothing better than terracotta pots or recycled tin cans (and with time and no small amount of skill, you could even make your own terracotta from the soil beneath your feet). If you have larger plants that need storing or growing on for later use, then a nursery bed can be constructed and the plants 'heeled in' (dug into the ground) until they need to be moved. Alternatively, hessian fabric is a wonderful way to hold soil around roots – tied in place with rope or twine – with the obvious benefit of being compostable after use. Elsewhere in the garden, plastic water cans can be replaced by reclaimed metal ones, plastic water butts can be replaced by old whisky barrels, wheelbarrows can be made from recycled pallets, and hand tools made from replaceable wooden handles.

Antiquities Mortar

As detailed earlier in the chapter, there will probably come a time when you will need to apply an extra layer of reinforcement to structures in the garden. This might be to strengthen a retaining wall, seal a firepit or oven, bolster load-bearing poles, or set hard-standing flagstones. To achieve this and avoid all inorganic

UPCYCLING TIN CANS FOR POTS

Tin cans are an abundant product with many upcycling possibilities. The quantity that my family goes through is immense, and although they are widely recycled, we must remember that the more we can do ourselves with the materials we acquire, the less energy is expelled in recycling them commercially.

1. Acquire some industrial-sized tin cans from schools, restaurants, cafés, etc. and save some small domestic tins.
2. Use a metal drill bit to drill a series of drainage holes in the bottom of the tin cans (figure 1).
3. Place stones, crocks or pebbles in the bottom, then fill with compost and plant up.

This recycled industrial baked bean tin can is perfect for a fast-growing mint.

Growing seeds and seedlings doesn't need to involve plastic and aesthetic uniformity.

MAKING MUD MORTARS

Although making bricks and pottery requires a certain degree of skill beyond that of the average gardener, making mud mortars does not. The raw materials for making a mortar lie beneath your feet.

1. Dig down to the subsoil to look for clay. If none is present, acquire some clay - sometimes for free - from building sites, recycling sites (figure 1), or upturned roots of fallen trees or riverbanks.
2. On boards, work the clay with your feet, removing large stones as you go, and adding water until the clay is sticky and malleable (figure 2).
3. Begin adding as much sand as the mix needs to reach a 50/50 proportion of clay and sand.

4 Using the mix as a mortar, point the stone or bricks (figure 3), keeping it as wet as possible to stop it from drying out too quickly.

compounds like concrete and cement, you'll need to make a clay or cob mortar. These mud mortars have their origins in the late Neolithic times and were still being used in the ancient civilisations of antiquity – before there was lime, there was mud. Even right up until the Middle Ages these mortars were used extensively in more rural societies or if lime wasn't readily available. And they are still very much in use today, especially in countries with arid climates. Of course, mud and clay aren't just used for mortar: they also formed the first bricks and pottery, with the earliest bricks dried by the sun until the advent of firing through kilns and ovens.

You can adapt the mud mortar into something more dense and robust by adding just one more element – straw (or other fibrous material, like sheep's wool). This is called cob. The straw in cob works to bind the mix to give more tensile strength, meaning you can use it to make retaining walls, load-bearing walls, or plasters for cob ovens and cob stoves. This mix is the same as adobe, mud bricks, and wattle and daub, seen and used for many centuries throughout the world. Constructions using these methods are usually (but not always) found in the drier climates, because the bricks or plasters can be sun-baked and hardened dry with relative ease. One step beyond this would be extracting pure clay from the soil to make tiles, pots and bricks. This is quite a lengthy process, whereby the other elements of the soil – sand and stone – are removed through stages of purifying, straining and drying until a

Cob is a wonderfully adaptable material, both strong and malleable.

pure clay putty is all that is left behind. It would be advisable to search out soils with a much higher content of clay before attempting this.

Gardens will always need materials, whether for practical purposes like walls, or to terrace the land or create compost bays, or for more aesthetic ambitions like pots, trellises and detailed edging. Even if we create a garden free from any stone, wood, brick or metal, materials will still play their part in the tools we use and the products we may need to buy. Their providence, production and permanence all need scrutiny more than ever before. We must consider the effect they will have on the wildlife already in the garden and the ecological impact they made on their place of origin. We should be mindful of how long it will take for them to degrade back into the earth and whether this will regenerate and enrich the soils in the future. In bypassing the use of any 'new' materials, we are rejecting some of the most destructive practices of extraction from the natural world, which will ease the pressure and burden on the ailing environment. And by learning and harnessing crafts and methods that work holistically alongside nature, such as coppicing and mud mortars, we can follow a path that leads to a future of balance and harmony in our material quest.

To craft a garden within these parameters needn't come at the expense of style and personal expression, however, as gardens made from reclaimed, reused and recycled materials can be at the very pinnacle of style, form and design. Gardens of maturity and age are evocative places that transport you elsewhere, and they attain a sense of grounding and place within their surroundings, while every new garden waits to be aged and weathered to reach its culmination. Bringing this atmosphere into the garden can be liberating and full of expression and invention, with no diminished effect on the composition or quality of the garden as a result. Whether you seek materials for practical or aesthetic purposes, you must be proactive in searching for used materials. Ask yourself what it is you need, and then determine who might have it but *doesn't* want it? It could be a school with an over-abundance of large tin cans that you can repurpose as plant pots, or a local farm tearing down an old barn full of timbers and corrugated roofing. The internet, reclamation and salvage yards, and auctions will always be of value as places to find used materials. But this is still a world that chews up and spits out so much material waste that there is gold to be found for everybody, as long as you are willing to go out and look for it.

To know that our choices can have profound, lasting and positive changes to the natural world beyond the garden gate can be thrilling, emboldening and uplifting. And in consequence, to know our gardens can be constructed and composed while still being a sanctuary for local wildlife and ecology is immensely inspiring.

I sometimes imagine a garden hundreds of years from now, generations after the last owners have passed and moved on. Maybe the house is gone and the area is being taken back by the wilderness; the trees are reaching maturity and the wild herbs and bulbs have colonised the ground, and all the wild animals have returned. Nature has reasserted its dominance. And I ask myself, are the plastic pots and concrete footings still there? Is there still a polytunnel and are there plastic tool handles littered on the ground? People like to say that we, as a species, will leave a thin layer of plastic all across the globe, pressed and forged into geological time for future archaeologists to find and ask why. Well, they don't have to in our gardens; they don't even have to know we were there at all.

ACKNOWLEDGEMENTS

This book wouldn't have been possible without the help, support and guidance of my wife, Bryony. Writing a book when you have a young family is a journey full of ups and downs, with many sacrifices and hard decisions. Bryony has been there through every moment, and her amazing spirit and endless love and patience are the reasons this book made it to the finishing line. I am forever grateful and ready to give back all the time I have taken.

I am not a writer who shuts the door and needs to be alone, so my whole family has been involved in the process in one way or another, whether they knew it or not! So my two daughters, Amber and Freida, need acknowledging too for understanding when Daddy has had to work long days and not play or read stories as often as usual. This book is theirs as much as it's mine and I only hope to share the joys the book will bring after putting in the hard yards together. Here's to my girls, I love you all!

I must also mention that I am a true believer in the power of collaboration, and two people in particular stand out as giants in bringing this book together. Firstly, Jon Rae, my commissioning editor at Chelsea Green. Jon picked up my proposal and helped turn a wild idea into a clear vision. The core message of hope and possibility that runs through the book has Jon's spirit within it, and I know with certainty that I wouldn't have got this far without his steadfast belief and guidance. Secondly, Michael Metivier, my developmental editor at Chelsea Green. Michael's mastery of the English language is evidenced throughout the book, as he helped shape and mould the text into more erudite and concise forms. There is no doubt that your enjoyment of the book and assimilation of its information was infinitely heightened by Michael's superb editing skills.

Lastly, the following were kind enough to allow me take photographs on their property and I thank them for their generosity and kindness:

Marianne and Adam Johnson – Somerset, UK
Tristan Faith – Somerset, UK
Samantha Evans – Somerset, UK

RESOURCES

While writing this book, my mind has been at times electrified, inspired and lucid, while also being saturated, docile and empty. This is the journey an author undertakes. The saving grace on this roller coaster is that I was not alone. I had the wisdom, knowledge and energising spirit of many authors, thinkers and scholars swimming in my thoughts and guiding me to the end. I urge you to seek them out, so that you too can further your own journey of discovery through the ecological sphere and come out more hopeful and inspired towards the future than you were when you began.

Recommended Reading for Inspiration

The Garden Jungle: or Gardening to Save the Planet by Dave Goulson (Jonathan Cape, 2019)

The Hidden Life of Trees: What They Feel, How They Communicate – Discoveries from a Secret World by Peter Wohlleben (Greystone Books, 2016)

Life on Earth: A Natural History by David Attenborough (William Collins & Sons, 1979)

The One-Straw Revolution: An Introduction to Natural Farming by Masanobu Fukuoka (Rodale Press, 1978)

Teaming with Microbes: The Organic Gardener's Guide to the Soil Food Web by Jeff Lowenfels and Wayne Lewis (Timber Press, 2010)

This Changes Everything: Capitalism vs. the Climate by Naomi Klein (Simon & Schuster, 2014)

Wilding: The Return of Nature to a British Farm by Isabella Tree (Picador, 2018)

Recommended Reading for Reference

Collins Tree Guide by Owen Johnson and David More (HarperCollins Publishers, 2004)

Flora Britannica by Richard Mabey (Chatto & Windus, 1996)

Making a Wildflower Meadow by Pam Lewis (Frances Lincoln, 2003)

RSPB Handbook of Garden Wildlife by Peter Holden and Geoffrey Abbott (Christopher Helm, 2008)

Wild Flowers by Sarah Raven (Bloomsbury, 2011)

Recommended Websites

PEOPLE

Charles Dowding
https://charlesdowding.co.uk

The foremost advocate of no-dig gardening in the UK. He's also a composting master and you can check out his techniques through his wonderful YouTube channel, linked to the website.

Geoff Lawton
www.geofflawtononline.com

The de facto leader of the permaculture movement. A student of Australian permaculturist Bill Mollison, Geoff took over Mollison's institute and now runs a similar set-up at Zaytuna Farm in eastern Australia. Again, I enjoy his excellent YouTube channel, where he unfurls his vast knowledge on all things ecological.

ORGANISATIONS

The Food Forest Project
www.thefoodforestproject.org

A burgeoning collective, inspiring and educating the local community about the positive benefits of permaculture. They work out of my hometown and I've been witness to their galvanising energy and forward-thinking ecology-first projects.

Plantlife
www.plantlife.org.uk/uk

A growing force in the fight against the ecological decline of British flora and fauna. Their campaigns are well worth getting behind and their website has a fantastic database of indigenous plants.

Send a Cow
https://sendacow.org

A non-profit charity founded between Africa and the UK in the 1980s. Working closely with rural African communities, they helped establish the innovative 'keyhole garden' design mentioned in this book.

The Wildlife Trusts
www.wildlifetrusts.org

An independent charity, concerned with the protection of wildlife and the natural world. They do great work for me locally in establishing nature reserves that I love to explore and learn more of the natural world. Their website is a vast resource of information and guidance on indigenous plants and animals.

OTHER

The Nito Project
www.youtube.com/c/TheNitoProject

Great modern videos of all the skills needed to make cob, wattle and daub, clay plasters and more.

Online Atlas of the British and Irish Flora
www.brc.ac.uk/plantatlas

The best online database of indigenous British plants that I know of. It can be a bit clunky and old-school but the depth of scholarly research behind it is phenomenal.

Recommended Suppliers

Emorsgate Seeds
https://wildseed.co.uk

A top supplier of native British wildflower seeds. They have specialist collections from specific locations – helping to keep the local flora biodiversity alive. There's plenty of advice and guidance on the website too.

Niwaki
www.niwaki.com

A great collection of Japanese-style gardening tools. They do sharp, precision blades and the waterstones (mentioned in the Materials chapter) to maintain these to the highest standard. I get my secateurs, sickles, shears and knives from here.

The Scythe Shop
www.thescytheshop.co.uk

Run by author, farmer and editor Simon Fairlie, this Dorset-based company supplies the best Austrian-style scythes in the UK. The website has plenty of information on maintenance, with links to videos and books.

The Walled Garden at Mells
www.thewalledgardenatmells.co.uk

A superb garden and plant nursery based in Frome, Somerset. They're always looking to develop and encourage permaculture principles within their working framework.

Wells Reclamation Company
www.wellsreclamation.com

My local reclamation yard, with a great range of reclaimed and recycled timbers, stone, bricks, ironworks and pots. Worth a visit for the yard alone – a labyrinth of materials and collectables.

INDEX

Note: Page numbers in *italics* refer to photographs. Page numbers followed by *t* refer to tables.

ABOUT THE AUTHOR

Matt Rees-Warren is an ecological gardener, designer and writer. During 15 years of life in the 'outside', he's worked for the National Trust, been head gardener at Kilver Court Gardens, Somerset, had articles published in RHS *The Garden*, *Somerset Life* and *Country Gardener*, and designed gardens for private clients in and around the South West. Through his work, Matt has looked to illuminate, propose and develop ecological gardening methods and practices. As an advocate of organic gardening, permaculture, no-dig gardening and wildlife gardening, among other practices, his style and ethos reflect the changing relationship between ourselves and our gardens, and the natural world.